Water for urban areas: Challenges and perspectives

Edited by Juha I. Uitto and
Asit K. Biswas

TOKYO · NEW YORK · PARIS

© The United Nations University, 2000

The views expressed in this publication are those of the authors and do not necessarily reflect the views of the United Nations University.

United Nations University Press
The United Nations University, 53-70, Jingumae 5-chome, Shibuya-ku, Tokyo, 150-8925, Japan
Tel: +81-3-3499-2811   Fax: +81-3-3406-7345
E-mail: sales@hq.unu.edu
http://www.unu.edu

United Nations University Office in North America
2 United Nations Plaza, Room DC2-1462-70, New York, NY 10017, USA
Tel: +1-212-963-6387   Fax: +1-212-371-9454
E-mail: unuona@igc.apc.org

United Nations University Press is the publishing division of the United Nations University.

Cover design by Joyce C. Weston

Printed in the United States of America

UNUP-1024
ISBN 92-808-1024-3

Library of Congress Cataloging-in-Publication Data
Water for urban areas : challenges and perspectives / edited by Juha I. Uitto and Asit K. Biswas.
    p. cm.
  Papers presented at the sixth UNU Global Environmental Forum, held at the headquarters of the United Nations University on 25 June 1997. Includes bibliographical references and index.
    ISBN
  1. Municipal water supply-Congresses. 2. Urbanization-Congresses. 3. Twenty-first century-Congresses. 4. Water-supply-Management-Case studies-Congresses. I. Uitto, Juha I. II. Biswas, Asit K. III. UNU Global Environmental Forum (6th : 1997 : United Nations University)
TD345.W2625   2000
363.6'1'091732—dc21   99-050477

Contents

List of tables and figures   vii

Preface   xi
Juha I. Uitto and Asit K. Biswas

Foreword   xv
Abraham Besrat, Vice-Rector, United Nations University

1 Water for urban areas of the developing world in the twenty-first century   1
Asit K. Biswas

2 Water management in metropolitan Tokyo   24
Yutaka Takahasi

3 Water quality management issues in the Kansai Metropolitan Region   47
Masahisa Nakamura

Contents

4 Water management in mega-cities in India: Mumbai, Delhi, Calcutta, and Chennai 84
   Rajendra Sagane

5 Water supply and distribution in the metropolitan area of Mexico City 112
   Cecilia Tortajada-Quiroz

6 Wastewater management and reuse in mega-cities 135
   Takashi Asano

7 The role of the private sector in the provision of water and wastewater services in urban areas 156
   Walter Stottmann

8 Emergency water supply and disaster vulnerability 200
   Charles Scawthorn

9 Conclusions 226
   Juha I. Uitto and Asit K. Biswas

Contributors 231

Index 232

Tables and figures

Tables

1.1	World population growth: Observed and projected, 1800–2103	3
1.2	Increase in population by region, 1995–2030	4
1.3	Population of the world's 10 largest mega-cities, 1994 and 2015	5
1.4	Water service indicators for selected Asian cities	20
2.1	History of waterworks in Tokyo	25
2.2	Trends in Tokyo's water leakage rate, 1915–1995	28
2.3	Purification plants for Tokyo's water supply	33
2.4	Dams in the Tama River and Tone River systems	34
2.5	Features of Tokyo's water service, 1984/5–1993/4	35
2.6	Features of the water service in Japan's main cities, 1994	39
2.7	Features of the water service in various cities of the world	40
3.1	Population served by Lake Biwa water, 1994	49
4.1	Water supply in the four mega-cities of India, 1997	88
4.2	Proposed dams to augment Mumbai's water supply	95
4.3	Planned dams to augment Delhi's water supply	101
4.4	Greater Calcutta's sources of drinking water	102

List of tables and figures

4.5	Projected population and demand for water in Calcutta, 2000 and 2015	105
5.1	Characteristics of the Mexico City Metropolitan Zone and use of water supplied to Mexico City and the State of Mexico	118
5.2	The results of some leakage studies in Mexico, 1991	121
5.3	Tariffs on average consumption in Mexico City, 1996 and 1997	121
6.1	Summary of the major parameters used to characterize reclaimed wastewater quality	140
6.2	Overview of representative unit processes and operations used in wastewater reclamation	143
6.3	Categories of municipal wastewater reuse	145
6.4	Comparison of the World Health Organization's microbiological quality guidelines and criteria for irrigation and the state of California's current wastewater reclamation criteria	150
6.5	Comparison of tertiary treatment costs for water reuse	151
7.1	Allocation of key responsibilities for private participation options	161
7.2	Some private sector contracts in place in water and sanitation	161
7.3	Potential stakeholder issues and policy responses	180
7.4	Private sector options and objectives	185
7.5	Private sector options and enabling conditions	186
8.1	CCWD system qualitative reliability summary: Fire following M6.5 Concord Fault earthquake	209
8.2	Reliability analysis: San Francisco reclaimed water/fire protection dual-use addition	213
8.3	Summary of reliability aspects of case-study water supply system projects	223

Figures

1.1	Patterns of population growth in selected mega-cities, 1840–1995	6
1.2	Current and projected future water costs	14
1.3	Use of bottled water in India, 1992–1997	18
2.1	Developments in Tokyo's water service, 1900–1995	26
2.2	Growth in the average daily water supply volume, 1900–1995	27

3.1	The Lake Biwa–Yodo River–Osaka Bay area and part of Kansai Metropolitan Region	48
3.2	Water resource development projects in the Lake Biwa–Yodo River region	53
3.3	Allocation of water rights in the Lake Biwa–Yodo River region	54
3.4	Trends in municipal and industrial water use in Osaka City and Osaka Prefecture, 1965–1994	58
3.5	Osaka City and Osaka Prefecture industrial water supply systems	61
3.6	Wastewater systems in Shiga Prefecture	64
3.7	Water supply intakes and wastewater effluent discharge points along the Lake Biwa–Yodo River watercourse	67
3.8	Concentrations of organophosphoric acid triesters in the Yodo River	70
3.9	Distribution of agrochemical residues in the Lake Biwa–Yodo River water	71
3.10	Trends in Lake Biwa water quality: TP and COD, 1979–1996	74
3.11	Trends in algal bloom incidents, 1977–1996	76
4.1	India's mega-cities	86
4.2	Greater Mumbai's water sources	90
4.3	Delhi's water sources	97
4.4	Tapping points on the River Ganga for Calcutta Metropolitan Area	103
4.5	Chennai's water sources	107
5.1	Water use by sector in Mexico, 1995	113
5.2	The availability of water in Mexico, 1990–1995	114
5.3	The availability of treated water in Mexico, 1990–1995	115
6.1	The role of engineered treatment, reclamation, and reuse facilities in the cycling of water through the hydrological cycle	138
6.2	Comparison of the distribution of reclaimed water applications in California, Florida, and Japan	147
7.1	Range of options for private sector participation in the municipal water and wastewater sector	170
8.1	Generalized methodology for water system reliability analysis	202
8.2	Iteration alternatives for CCWD cost–benefit analysis	210

List of tables and figures

8.3	San Francisco Auxiliary Water Supply System, including dual-use addition in western portion of San Francisco	212
8.4	Dedicated Fire Protection System, City of Vancouver	215
8.5	Rural water purification distribution system	218
8.6	Incidence of cholera, dysentery, and dehydration in refugee camps in Goma, Zaire, July–August 1994	219
8.7	Daily mortality in refugee camps in Goma, Zaire, July–August 1994	219

Preface

The focus of the Sixth Global Environmental Forum, which was convened by the United Nations University (UNU) in Tokyo, Japan, on 25 June 1997, was on "Water for Urban Areas in the 21st Century." The sixth Forum, like the preceding five, was organized with the support of a leading Japanese construction company, Obayashi Corporation.

The topic for the Forum was selected because water and wastewater management for the urban areas of the developing world are likely to become increasingly important and complex tasks during the first half of the twenty-first century. When the issues of increasing water scarcities and accelerating water pollution in and around the urban centres of the developing world are superimposed on the continuing trend of rapid urbanization, the magnitude and the extent of the problems associated with the issue of water and wastewater management for the urban areas during the twenty-first century are likely to increase significantly compared with what they are at present.

The problems facing the urban areas of the developing world can be realized from the following selected facts:

Urbanization
- In 1950, 30 per cent of the global population lived in the urban areas. By 1995, this figure had increased to 45 per cent, a 50 per cent increase. In certain countries such as Nigeria, the urban population increased by more than four times during this period.
- By 1995, 70 per cent of the population of Europe, North America, Latin America, and the Caribbean were living in urban areas.
- Currently some 30–35 per cent of people in Africa and Asia live in urban areas. These two continents are now witnessing high urban growth rates estimated at roughly 4 per cent per year.
- Between 1950 and 1990, the number of cities with a population of over 1 million increased almost four-fold, from 78 to 290. By 2025, this number is estimated to increase to more than 600.
- By 2015, the population of mega-cities such as Bombay, Jakarta, Karachi, and Lagos is expected to increase by more than 90 per cent. In contrast, Tokyo is anticipated to grow by only 10 per cent during the same period.

Water supply and wastewater management
- In 1994, approximately 280 million people in urban areas did not have access to a water supply. Among the urban poor, less than 30 per cent of households were connected to water distribution systems.
- Very few urban centres can provide water that can be drunk safely. Not surprisingly, the consumption of bottled water among the affluent sections of society is increasing exponentially. In India, for example, between only 1992 and 1997, the annual consumption of bottled water increased 4.5 times.
- In 1990, 453 million people (that is, 33 per cent of the urban population) had no access to sanitation. By 1994, this number had increased to 589 million people, or 37 per cent of the urban population. Present trends indicate further increases in coming years.
- Irrespective of official statements, only about 2–6 per cent of sewage collected in major urban centres of Latin America is properly treated at present.
- The costs per cubic metre of water for new water supply projects in real terms are now 1.5 to 3 times the cost of the previous generation of projects. Thus, the investment requirements for new water projects would be significantly higher than are estimated at present.

Preface

Health impacts of the above factors
- Unsafe drinking water is responsible for 80 per cent of diseases and 30 per cent of deaths in developing countries.
- Altogether 1.2 billion people suffer annually from diseases caused by unsafe drinking water and/or poor sanitation.
- More than 4 million children die each year from waterborne diseases.
- In the developing world, 15 per cent of children will die before reaching the age of 5 as a result of diarrhoea, which is caused by poor water supply and sanitation.

The above facts, as well as numerous other associated issues, clearly indicate that the developing world faces a mammoth problem in coming decades, in terms of water supply and wastewater management in urban areas, whose magnitude and complexity no other generation in human history has had to face. In addition, and irrespective of global rhetoric, the overall global situation in terms of good-quality drinking water and proper wastewater treatment during the 1990s has been progressively deteriorating, in terms of both absolute numbers as well as the percentage of the urban population affected.

Because of this critical situation and current unsatisfactory global trends, the United Nations University decided that the Sixth Global Environmental Forum would focus on water for urban areas in the twenty-first century. Eight of the world's leading experts from different disciplines, institutions, and countries were specifically selected for the Forum, and were then invited to prepare background papers within an overall integrated framework, and also to lead the discussions during the Forum. We are indeed very pleased that all our first-choice speakers promptly agreed to participate in this important meeting. The Forum was specifically designed to be a very focused event, and the various complex and interrelated problems were addressed from multidisciplinary and multisectoral viewpoints as well as regional perspectives.

The Forum was convened at the headquarters of the United Nations University, and some 350 participants from different institutions all over Japan and from international institutions active in Japan attended the event. Following the presentations of the invited world experts, there was a Panel Discussion with extensive audience participation.

Preface

We are most grateful to the authors of the background papers for accepting our invitation, as well as to Obayashi Corporation for their generous financial sponsorship that made the Forum possible.

Juha I. Uitto
Asit K. Biswas

Foreword

Abraham Besrat, Vice-Rector, United Nations University

The Sixth UNU Global Environmental Forum, "Water for Urban Areas in the 21st Century," was organized by the United Nations University and sponsored by Obayashi Corporation on 25 June 1997. Both topics – water and cities – will be critical in the next century.

The past several decades have seen the proliferation of cities and urban areas around the world. It is estimated that, by the end of the twentieth century, more than half of the world population will live in cities, and, by the year 2025, more than two-thirds of the population will do so, with the highest urban growth rates in Asia and sub-Saharan Africa. Such growth will bring tremendous stress and pose formidable problems of social and institutional change, infrastructure investment, and pollution control. Even today, taking into account the recent growth in environmental infrastructure investments worldwide, the World Bank estimates that about 380 million urban residents still do not have adequate sanitation, and at least 170 million still lack access to a nearby source of safe drinking water. This problem does not preclude Japan and other developed countries. Urban population growth is universal on this planet, and virtually all cities are experiencing environmental degradation.

Foreword

One very important problem that we will face in the twenty-first century is water: water that is needed to support our cities; water that we can drink; water that we can cook with; water that we can use for cleaning; water that we can use in our industries. Making clean water available in the next 40 or so years will require extending the service to 3.7 billion more urban residents. In addition, preventing further degradation in fast-growing countries will require that pollution per unit of industrial output fall by 90 per cent between now and the year 2030. This is a difficult task indeed, considering that most developing countries today would put development before environmental concerns.

The problems related to water are manifold and contrasting. Either there is too little of it – as during a drought or an extremely hot summer – or there is too much of it – as during flooding. Aside from the problem of quantity, there is also a problem of quality: Is the water safe to drink? Is the taste acceptable? Are we recycling it?

For a developing country city, the most immediate problems are the health impacts of urban pollution that derive from inadequate water, sanitation, drainage, and solid waste services. Together with poor industrial waste management and air pollution, these problems combine into what is called the "brown agenda" – a set of problems linked to poverty and environment.

Because poverty, economic development, and the environment are inextricably linked, another problem that we can foresee in the very near future is the issue of equity. This means different income brackets getting a different quantity and quality of water and services.

We mostly still consider water a free resource. The thinking that anything that falls from the sky is free must change very soon. Recent studies have shown that water prices will double, if not triple, in the next few years because of the commodity's increasing scarcity in a state that we can use it. This impending change, and others that go with it, will disproportionately affect the urban poor, and their plight in turn will exacerbate the urban environmental crises. According to one estimate, the cost of pollution problems alone in developing countries exceeds 5 per cent of their GDP. Clearly, improving the situation of the urban poor is an essential precondition for reducing urban environmental hazards.

The United Nations policy on water is guided by the Agenda 21 adopted at the United Nations Conference on Environment and Development (UNCED), the so-called Earth Summit, organized in Rio de Janeiro in 1992. Although only one chapter is directly concerned

Foreword

with the protection of the quality and supply of freshwater resources, virtually all other chapters that deal with the conservation and management of resources for development also deal with water issues.

Sustainable management of water resources also figures centrally in the United Nations University's research and training programme. The current programme includes projects on such topics as hydropolitics and eco-political decision-making, water quality assessment and monitoring, and environmental governance. The focus of the UNU's work is on developing countries in the Middle East, Africa, Asia, and Latin America. Only recently, the University inaugurated a new research and training programme – the International Network on Water, Environment and Health (UNU/INWEH). The aim of UNU/INWEH is to link the watershed-ecosystem approach with human health needs, giving full consideration to the developmental aspects of water, which is so essential for human life and work. The UNU/INWEH programme will focus on practical problem-solving, with particular emphasis on the needs and concerns of the developing regions of the world.

In June 1997, world leaders again gathered together under the United Nations umbrella to discuss the progress made and the problems faced in the five years following the Rio summit. The United Nations General Assembly Special Session Earth Summit+5 in New York considered freshwater to be one of the central issues facing humankind as we strive towards sustainable development.

Also in 1997, the Denver Summit of Eight, which brought together the heads of the leading nations, highlighted the need to promote sustainable development and protect the environment. One of the issues singled out in their final communiqué was freshwater; they called for the promotion of efficient water use, improvement of water quality and sanitation, technological development and capacity building, public awareness, and institutional improvements.

Japan, as one of the leading nations in both financial and know-how terms, is well placed to take a lead in this process. In his speech to the UN General Assembly Special Session, Prime Minister Ryutaro Hashimoto emphasized the role Japan could have in promoting the development and spread of environmentally sound technologies.

The attention paid to freshwater by the world community demonstrates the gravity of the issues involved. The Sixth UNU Global Environmental Forum brought together leading experts from around the world to discuss solutions to the problems related to providing adequate and clean water supply and sanitation for the growing ur-

Foreword

ban centres. The Forum considered mega-cities of both industrialized and developing nations, and covered topics such as wastewater management and reuse, the role of the private sector, and emergency water supply and disaster vulnerability.

The Forum exemplified the mode of operation of the UNU in disseminating state-of-the-art knowledge on pressing global problems. An autonomous academic organization under the United Nations umbrella, the UNU brings together networks of eminent scholars and experts to work on important issues that are the concern of the world body and to present their research results and experience to the public through forums such as this. This Forum was the sixth annual UNU Global Environmental Forum organized since the series began in 1991. The purpose of these forums is to highlight contemporary environmental issues and to disseminate research results to the public, thereby creating greater awareness of the challenges that lie ahead of us.

The United Nations University would like to thank Obayashi Corporation for their generous sponsorship of this Forum, as well as the ones that have taken place before it. This is, indeed, a very fine example of the public and private sectors cooperating and working together towards a better world.

1

Water for urban areas of the developing world in the twenty-first century

Asit K. Biswas

Introduction

Historically and culturally, water has always been considered to be a critically important resource, because without it ecosystems cannot survive. Similarly, human evolution simply would not have been possible without water. Thus, not surprisingly, water has a special place in the human psyche. For example, all major religions such as Christianity, Hinduism, and Islam give special importance and consideration to water. People, irrespective of which country of the world they live in, are generally against large-scale transfer of water to another country, whereas international trade in all other resources, both renewable and non-renewable, has been a routine activity for centuries and has seldom attracted protracted national discussions or serious political opposition. In contrast, water transfer plans between countries, and often between states within the same country, attract considerable political controversy.

From the dawn of history, water has continued to be an economically essential resource for the development of countries. However, during the twentieth century, especially during the latter half, all the western economies became stronger and more resilient than ever before, and became less and less dependent on water and the vagaries of nature. The easy availability of clean water for drinking twenty-four hours a day, at affordable prices, became the norm rather than an exception. Agriculture and food production became surplus to the needs of the people of the western world, and the vast majority of the important hydropower plants have now been constructed. Floods and droughts, for the most part, are already controlled through extensive water development projects. Concerns are now expressed only when there are catastrophic floods or prolonged severe droughts, both of which are temporary in nature. Public, political, and media interest in water, which is extensive during catastrophic events, mostly evaporates after the floods and droughts are gone, and reappears only after several years or even decades, when the next catastrophe strikes. Accordingly there is little sustained public and political interest in the developed countries in water. The availability of clean water or the impact of water availability on society are no longer major critical issues for the developed world. This situation is unlikely to change in the twenty-first century.

In contrast, the situation is very different from the perspectives of the developing countries. Availability of clean water is still a dream in most parts of the developing world. For mega-cities, from Calcutta to Istanbul, and Mexico City to Lagos, clean drinking water is simply not available from the urban water supply systems. For health reasons, drinking water has to be either boiled or filtered, or has to be consumed from bottles. Not surprisingly, the failure of the governmental system to provide clean drinking water to the citizens of Mexico City has been very efficiently compensated for by the private sector, which has developed a sophisticated and extensive distribution system for drinking water in very large plastic bottles that compares most favourably with any distribution system available for soft drinks anywhere in the world. Clean drinking water is supplied, regularly and cost-effectively, to all the households of the city that can afford it. The empty bottles are collected for reuse, along with deliveries of new supplies.

In addition to the lack of safe drinking water in the urban areas of the developing world, the developing economies are still very much dependent on water availability because of its importance to agricultural production, industrial development, and electricity generation. With expanding population and urbanization, accelerating human activities, and increasing per capita water consumption, an adequate supply of water of the right quality and subsequent wastewater disposal in an environmentally acceptable way are likely to become even more important to the developing world during the next two to three decades.

Population and water

From the dawn of history, as human population has continuously increased, so too have water and wastewater disposal requirements. The most important early civilizations developed along the banks of major rivers such as the Nile, the Euphrates–Tigris, and the Indus. These rivers provided water not only for drinking but also for irrigation.

Water availability was generally not a serious problem as long as the population numbers were low and the concentrations of people were not high. As the population started to increase dramatically during the post-1950 period, and the rate of urbanization began to accelerate, the provision of clean water and the environmentally sound disposal of wastewater for the urban areas of developing countries became serious problems. These problems have become increasingly bigger and more complex during the past two decades. This trend is likely to continue, at least for another two to three decades. The immensity of the problem caused by the global population increase can be seen in table 1.1.

The global population is currently estimated to double between 1990 and 2100, but most of this increase is likely to occur by the year 2025. The population of the present low-income countries is expected to increase by 235 per cent; in contrast, the corresponding increase in

Table 1.1 **World population growth: Observed and projected, 1800–2103**

Year	Population (billion)	Years to next billion
1800	1	125
1925	2	35
1960	3	14
1974	4	13
1987	5	12
1999	6	12
2011	7	12
2023	8	16
2039	9	21
2060	10	43
2103	11	–

Source: Bos et al. (1994).

Table 1.2 **Increase in population by region, 1995–2030**

Region	Population (million)		Percentage increase
	1995	2030	
Africa	720	1,600	116
Asia	3,400	5,100	47
Europe	731	742	1
Latin America and Caribbean	475	715	51
North America	295	368	24
Oceania	29	39	36

Source: Bos et al. (1994).

high-income countries is likely to be less than 10 per cent. Continent-wise, Africa is projected to grow the most, more than four-fold during this period. In contrast, the population of Europe in 2100 is expected to be less than it is at present. Table 1.2 shows population growth in the various regions of the world between 1995 and 2030.

Urbanization and water

The problem of providing clean water and sanitation to the world's steadily expanding population has been worsened by the rapid urbanization process in the developing countries from around the middle of the twentieth century. According to an estimate by the United Nations, nearly 45 per cent of the total population of the world lived in urban areas in 1995; the corresponding estimate for 1950 was only 30 per cent. The global figures are averages: they mask wide disparities from one part of the world to another. For example, in the case of Nigeria, less than 10 per cent of the people lived in urban areas in 1950. By 1991, the proportion had risen to 42 per cent, and the trend has been accelerating during the past decade (Okunlola, 1996).

Although the mega-cities of the developing world have attracted the most attention from various international organizations in recent years in terms of the provision of adequate water supply and sanitation services, it should be noted that they account for a very small percentage of the global population. This holds true even though they consume the lion's share of national resources in terms of the necessary infrastructure development and management. If mega-cities are defined to be those having more than 10 million residents, only about

Table 1.3 **Population of the world's 10 largest mega-cities, 1994 and 2015**

1994		2015	
City	Population (million)	City	Population (million)
Tokyo	26.5	Tokyo	28.7
New York	16.3	Bombay	27.4
São Paulo	16.1	Lagos	24.4
Mexico City	15.5	Shanghai	23.4
Shanghai	14.5	Jakarta	21.2
Bombay	14.2	São Paulo	20.8
Los Angeles	12.2	Karachi	20.6
Beijing	12.0	Beijing	19.4
Calcutta	11.8	Dhaka	19.0
Seoul	11.5	Mexico City	18.9

3 per cent of the global population lived in such large urban agglomerations in 1990. While the mega-cities present tremendous management challenges at present, and will continue to do so in the foreseeable future, much of the recent urban growth is being witnessed in medium to small urban centres in many developing countries. This trend is likely to continue in the coming decades, and may even accelerate. This issue will be discussed further later in this paper.

The rapid growth of some of the mega-cities of the developing world will undoubtedly pose major planning and management challenges in the twenty-first century. In 1994, of the world's 10 largest mega-cities, 3 were in the developed world. By 2015, only one, Tokyo, is expected to remain on that list (table 1.3). Even though Tokyo is likely to continue as the largest mega-city, its population is estimated to increase by less than 10 per cent. In contrast, cities such as Bombay, Lagos, Jakarta, or Karachi are expected to grow by more than 90 per cent. Managing such high growth rates within a short period of only two decades efficiently and equitably would be a most difficult task in the best of circumstances.

It should be noted that urbanization and the formation of mega-cities are not new phenomena; cities such as London or New York started to grow in the nineteenth century (fig. 1.1). However, two major factors should be noted that have made the urbanization process and provision of water supply and sanitation services in the mega-cities of the developed world fundamentally very different from their counterparts in developing countries nearly one century later.

The first factor is the rate of growth. The development of mega-

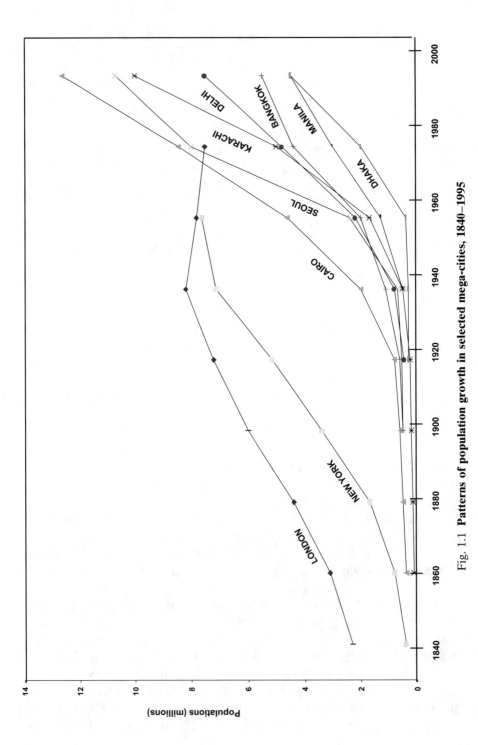

Fig. 1.1 **Patterns of population growth in selected mega-cities, 1840–1995**

cities in the developed world was a gradual process. For example, most of the population growth in cities such as London and New York was spread over a century (fig. 1.1). This gradual growth rate enabled these cities progressively and effectively to develop the necessary infrastructures and the capacities to manage their water supply and sewerage services. It was not an easy task but it was manageable.

In contrast, most of the urbanization in the cities of the developing world (e.g. Cairo, Seoul, Manila, and Karachi) occurred during the post-1950 period, and the really explosive growth generally took place after 1960 (fig. 1.1). The mega-cities and other major urban centres of the developing world simply could not cope with these very high and continually increasing urbanization processes; they were unprepared to manage such explosive growth. Thus, the quality of life declined rapidly during such high urbanization periods. For example, between 1950 and 1980, the population of Mexico City increased from 3.137 million to 13.354 million, a 425 per cent increase. To a certain extent, many of these mega-cities could handle the provision of water supply, but they generally fell progressively behind in constructing and properly managing sewage and wastewater treatment facilities. Even in a region such as Latin America, where many cities made reasonably good progress in installing sewerage systems, concomitant progress did not occur in wastewater treatment. Currently, only about 2–6 per cent of sewage collected in major urban centres of Latin America receives adequate treatment. Thus, in major cities ranging from Bogotá to Buenos Aires, and Mexico City to Santiago, some 50–60 million m^3 of mostly untreated sewage are discharged daily into nearby water bodies. Many of the governments claim that 10–20 per cent of wastewater is treated, but these are inflated figures; they include many treatment plants that are no longer operational, and also wastewater that receives inadequate treatment.

The second major factor is that, as the urban centres of the industrialized countries expanded, their economies were growing concomitantly. Accordingly, these centres were economically able to harness resources to provide their citizens with appropriate water supply and sewerage services. Even a country like Japan could invest heavily in the construction of urban infrastructures, including water supply, sewerage, and flood drainage services, after the Second World War because its economy continued to expand significantly during this period as well. Such extensive infrastructure development and major improvements in management practices meant that water losses due to leakages from urban water supply systems could be reduced drastically from a post-war estimate of 80 per cent to about 8–10 per cent, which is one of the lowest losses encountered anywhere in the world at present.

In stark contrast to the above, during the past three decades economies of the developing world have not performed very well. High public debts and inefficient resource allocations have ensured that the investments needed to construct all types of new urban infrastructures and maintain the existing ones have not been forthcoming. Lack of proper planning and poor management practices have further aggravated the situation, especially in terms of the proper treatment of drinking water and the disposal of wastewater.

Although considerable progress has been made in recent years in providing drinking water in urban areas, commensurate advances in sanitation have been for the most part missing in much of the developing world. For example, current estimates indicate that 453 million people (that is, 33 per cent of the urban population) had no access to sanitation in 1990. By 1994, this number had increased to 589 million (37 per cent of the urban population). Current estimates indicate that the number of people not having adequate sanitation has increased to about 850 million (approximately 50 per cent of the population). It should be noted that all these figures are gross estimates. However, they do give an indication of the magnitude of the problem the developing world is now facing, which is one that needs to be resolved as soon as possible.

Continuing urbanization poses a major challenge in terms of the provision of water supply and sanitation services, but its importance and contribution towards the development of stronger and more stable national economies should not be underestimated. It has been estimated that the urban areas of the developing world, which contain about 30 per cent of the total population, contribute nearly 60 per cent of the total gross national product, and also play an equally important and prominent role in terms of social development and cultural enhancement. Thus, the urbanization process presents both challenges as well as opportunities.

The main problems of major urban centres often stem from the fact that the rates of urbanization have often far exceeded the capacities of the national and the local governments to plan and manage the demographic transition efficiently, equitably, and sustainably, and to provide and maintain the necessary infrastructural development, services, and employment. The accelerating urban growth rates have mostly overwhelmed the limited capacities and resources of governments at all levels. Unplanned and poorly managed urbanization processes have unquestionably been an important source of social and environmental stress in all developing countries. The effects have been manifested in extensive air, water, land, and noise pollution, which have and will continue to have major impacts on the health and welfare of urban dwellers in the developing world for many years to come. The problem has been further compounded by increasingly skewed income distribution, which is continuing to worsen with time, high rates of unemployment as well as underemployment, corruption at all levels, and high crime rates.

A 1988 report from the United Nations summed up the grim situation for Karachi, a major Asian urban agglomeration. The situation is somewhat similar for most other mega-cities of Asia, Africa, and Latin America. The report (United Nations, 1988) stated:

Karachi, which has a high rate of natural increase and continuing influx of migrants from upcountry, is currently one of the most rapidly growing mega-cities in the developing world. Because Karachi is growing by more than five percent per annum, many basic services are strained to the point of collapse. Moreover, much of Karachi's population increase is being accommodated in *katcchi abadis* – sprawling, unserviced squatter settlements that have become breeding grounds for social unrest.

The two main problems faced by the major urban centres often further complexify an already difficult and complex situation. The first of these problems is the sudden fast rate of growth, often after decades of primarily horizontal expansion, especially in the central business areas. This contributes to a sudden surge in population densities, with concomitant high water requirements and the generation of high waste loads per unit area. The existing water supply and sanitation services and the poor planning capabilities of the authorities are mostly unable to cope successfully with almost instantaneous growths in demand for water and sanitation services.

The overall problem of the mega-cities is further compounded by the presence of informal and squatter settlements. Such settlements often account for 30–60 per cent of the total urban population. For example, it is estimated that approximately half of Bombay's population lives in such squatter areas. These areas are highly congested, leaving very little or no space for the provision of in-house water supply and/or public sanitation facilities.

The situation is further compounded by the fact that all levels of government generally give lower priorities to informal settlements. Resources are normally channelled to the areas where rich and important people live. In addition, urban planners often believe that adequate cost recovery for the provision of services to these settlements is not possible, because they are inhabited by poor or very poor people. Accordingly, informal settlements are often neglected or receive much lower priority in terms of the allocation of resources, and thus services, compared with rich and middle-class areas. Furthermore, the population of these informal settlements continues to grow owing to the regular arrival of rural people, who migrate to the urban areas in search of a better quality of life. Thus, whatever limited services may be available in the informal settlements become overwhelmed with the arrival of new people, and the limited water supply and sanitation services become progressively less and less adequate for serving an increasing population. This contributes to a progressive reduction in services that were inadequate to start with, and this deterioration in turn further increases the environmental and health risks to the people living in such areas.

Constraints on water availability

The provision of clean drinking water to the rapidly growing centres of the developing world and the safe disposal of wastewater face numerous constraints, which are complex and are often interrelated. Very few resolutions or declarations at various international forums, such as the United Nations General Assembly resolution declaring the decade of the 1980s as the International Water Supply and Sanitation Decade (IWSSD), have had any visible and perceptible impacts. It was evident in the 1970s that the goals set for this Decade, however laudable, were impossible to achieve unless major structural changes could be made in terms of resource allocation to the sector, both nationally and internationally. Not surprisingly, these goals were not achieved and indeed fell considerably short of expectations. However, the Decade did manage to put water supply and sanitation issues somewhat higher up the agendas of many governments of developing countries and external support agencies, especially towards the beginning and end of this period. Unfortunately, no real evaluation was carried out as to the actual impacts of the Decade per se, and thus the lessons that could have been learned from the Decade experiences are not known. For example, would the global situation been different had there been no Decade? If so, how would it have been different, by how much, and why? Anecdotal evidence indicates that developing countries and many external support agencies were already becoming aware of the importance of water supply and sanitation issues, and many of the developments that have occurred since 1980 probably would, in all likelihood, have occurred even without the Decade. The Decade, however, provided a strong focus, and some countries clearly took advantage of this fact to accelerate their programme activities in this area. Thus, overall, the Decade has probably contributed to a modest advance in access to water supply and sanitation services, which otherwise might not have occurred.

There are numerous major constraints that have to be overcome simultaneously before everyone in developing countries can have access to clean drinking water and sanitation services. Overcoming these constraints will not be an easy task, nor is it likely to be achieved within the next decade or so, irrespective of the numerous resolutions at organizations like the United Nations, or at many other major international forums. However, the future is not completely bleak, and there are some positive and encouraging signs. For example, in coming decades, as the population growth rates in developing countries continue gradually to decline and developing countries approach stationary populations, they are likely to become better positioned to be able to provide clean water and sanitation services to all their citizens. Population stabilization would mean that services would not continually have to cater to a moving target, as is the case at present.

Only some of the major constraints to achieving the ambitious goal of providing clean water and sewerage services to everyone in the world can be discussed briefly here because of lack of time and space.

Water scarcity

Water scarcity presents both a challenge and an opportunity in terms of urban water supply and sanitation. It is a challenge because new sources of water that could be developed cost-effectively are simply not available for most of the major urban areas of the developing world. From Amman to Beijing, and from Madras to Mexico City, there are simply no new sources of water that can be harnessed economically to quench the continually increasing urban-industrial thirst.

Because the existing sources of water that could be developed cost-effectively have already been developed or are in the process of development, and water that has been harnessed has already been fully allocated and in many cases over-allocated, additional supplies of drinking water can be obtained only by reallocating water that is currently used by other sectors, especially agriculture. National policies, explicitly or implicitly, generally give the highest priority to the domestic sector amongst all uses. However, it would not be an easy task, socially and politically, to transfer water from the agricultural to the domestic sector continually. Some such transfers can already be noted, but they are not widespread owing to deliberate policy decisions.

A good example is water availability in the Chennai (formerly Madras) area. In order to improve water-use efficiencies, the canals are being lined. The final result is really a *de facto* transfer of water from the agricultural to the domestic sector. Such practices are exacerbating the conflicts between agricultural and urban water users around many major cities all over the developing world, ranging from Manila to Mexico City (Naranjo Pérez de León and Biswas, 1997). Such conflicts are likely to increase significantly in the future, in terms of both intensity as well as geographical distribution, as various interest groups require more and more water.

Urbanization also brings opportunities. Urban centres are now important users of water. However, urban use takes place within a limited geographical area. Since domestic uses do not contribute to the actual consumption of water, all the water being supplied to households can be recaptured as wastewater through adequate sewerage networks. If this wastewater is then appropriately treated, it could serve as a "new" source of water. Although this treated wastewater may be restricted to specific types of water use owing to quality considerations and cultural reasons, it could still be used for certain purposes, thereby releasing higher-quality water for those uses that warrant it. The marginal cost of providing additional good-quality water in the same volume as treated wastewater would generally be much higher (Biswas and Arar, 1988), and the time required to obtain additional new good-quality water would be much longer.

Wastewater will be produced in urban areas irrespective of whether or not it is made use of. It is therefore essential that wastewater be treated adequately so as to reduce the environmental and health hazards for people living in and around the urban areas, which currently use existing unsatisfactory wastewater disposal practices in most developing countries.

Thus, increasing water scarcity can in one sense be considered an opportunity for encouraging urban areas to collect and treat wastewater properly so that it can subsequently be used as a "new" source of water to alleviate the scarcity.

High economic costs

Economic factors are becoming an increasingly important consideration in the provision of water supply and sanitation to the urban areas of developing countries.

For much of the developing world, for the most part, all the easily exploitable sources of water have already been developed, or are currently in the process of being developed. This means that the water sources that are yet to be developed are geographically, technologically, and environmentally more complex to handle, and accordingly the cost of harnessing and bringing this new water to the urban areas is now very high in real terms, especially compared with the cost of the earlier or present generation of water projects. For example, the average cost of providing storage for each cubic metre of river flow in Japan has increased nearly four-fold during the past 10 years (Biswas, 1997a). Approximately 20 per cent of this additional cost can be attributed to new social and environmental requirements. The major part of the additional cost, around 80 per cent, is due to the fact that the new projects are inherently more complex to construct, and are often located in more inhospitable terrain. Accordingly, the construction costs of these new projects are significantly higher when compared with already completed projects.

The current situation in terms of the costs of construction of new projects for supplying additional water to the major urban areas of the developing world is somewhat similar to that of Japan. For example, the World Bank's (1992) analyses of domestic water supply projects from various developing countries indicate that the cost of development of each cubic metre of water for the next generation of projects is often two to three times higher than that of the present generation. Figure 1.2 shows the current and projected costs in 1988 constant dollars for supplying a cubic metre of water to many major urban centres of the developing world.

In the area of wastewater treatment, the situation is no better. Most of the wastewater produced in urban areas is either not treated at all or receives inadequate treatment. Many governments, ranging from Egypt to Mexico, have in the past adopted the water quality standards of the developed world for internal and/or external politi-

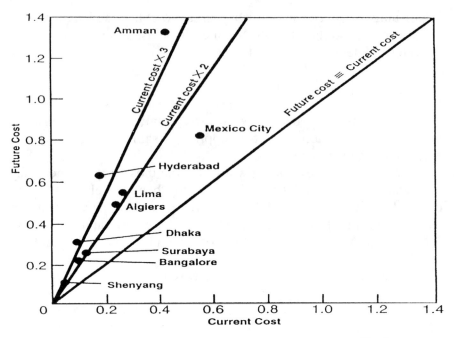

Fig. 1.2 **Current and projected future water costs (1988 US$/m^3)**

cal reasons, and because of faulty and incomplete analyses. No consideration was generally given to the fact that the standards that are appropriate and implementable for London and New York may be irrelevant, unimplementable, and often even counterproductive for Lima or Yaounde. Equally, no serious analyses are generally carried out into whether or not the standards adopted are essential for health reasons, or whether or not the countries concerned have the necessary financial resources and management capabilities to implement the stipulated standards (Biswas, 1997b). Not surprisingly, the promulgation of such inappropriate standards has generally not even helped to maintain, let alone improve, the somewhat poor quality of receiving waters in and around urban areas. The existing philosophies behind the formulation of water quality standards need to be carefully examined, especially as regards their implementation.

Financing and financial management constraints

The availability of adequate funds and the release of the funds in a timely manner to operate and maintain existing facilities are generally major constraints. Water utilities in developing countries are at present predominantly in the public sector, though private sector involvement is now being considered in one form or another in many parts of the world. The operation and maintenance of existing water supply and waste treatment systems, as well as the construction of new systems, are thus often constrained by lack of funds.

The economic situation is further compounded by inadequate pricing and inefficient billing and bill collection systems in most urban utilities. An analysis of 50 water utilities and bill collection systems in 31 Asian countries (McIntosh and Yñiguez, 1997) indicated the following major shortcomings:

- Fewer than 50 per cent of connections are metered properly. Some major cities, such as Calcutta, have no metering at all, and many have very little metering. Regular monitoring and replacement of faulty meters are exceptions rather than norms.
- Monthly household water bills in many major cities, such as Beijing, Tianjin, Hanoi, Mumbai (formerly Bombay), and Tashkent, are less than US$1.00. In contrast, the average monthly bill for well-managed utilities such as Hong Kong is US$31.00. Very low monthly bills encourage extravagant consumption and high wastage rates. Electricity to water bill ratios now average 18.5 for Faisalabad, 12.7 for Karachi, 9.2 for Tashkent, 7.8 for Kathmandu, and 7.7 for Delhi. Ratios of more than 4.0 generally indicate low water tariffs.
- The financial management of many utilities leaves much to be desired. For example, accounts receivable should be less than the equivalent of three months' sales. However, they are currently 19.7 months for Mumbai, 16.8 months for Karachi, and around 11 months for Dhaka and Shanghai.
- The utilities have different concepts of what constitutes operation and management expenses. Many normal operation and maintenance expenditures are left to be rectified by new investment projects. Such expenditures include the replacement of pipes, valves, water meters, and service vehicles, and the reduction of unaccounted for water. Thus, major investments are made in constructing new systems that are not properly maintained. System inefficiencies steadily increase owing to continuing deterioration, and then new investment projects have to be undertaken to rehabilitate the badly managed systems. Not only is this an inefficient use of capital but system efficiencies start to decline steadily from inception because of poor operation and maintenance practices. During

the entire process, the customers of the water utilities receive a poor service.
- Utilities are often over-staffed, and the staff are for the most part not properly trained and are inefficiently used. This generally contributes to low financial returns. For example, the number of staff per 1,000 connections is only two for a well-managed utility like Singapore, but very high for Tianjin (49.9), Mumbai (33.3), Beijing (27.2), and Chennai (25.9). The high ratios indicate poor efficiency. Low staff ratios could also indicate that many services are being contracted out to the private sector.

Management constraints

One of the major reasons for the poor performance and/or efficiencies of water utilities is their poor management. The poor management stems from two factors: unattractive salaries and regular political interference in management practices and decision-making processes.

In many urban areas, management compensation rates are determined by government salaries, because the utilities are in the public sector and thus must follow public service rules. Since private sector salaries are much higher than their public sector counterparts, bright and competent managers generally tend to gravitate towards private sector enterprises, where, in addition, there is also much less day-to-day interference from politicians. Politicians need to appreciate that multi-million dollar water and wastewater utilities cannot be managed efficiently by unqualified and inexperienced managers, with continual political interference, ranging from the recruitment of staff to even making some of the mundane decisions.

Analyses of the compensation packages of water managers in the Asian developing countries show very wide variances. For some utilities, annual salaries are less than US$1,000, but they can be as high as US$145,000. Not surprisingly, the well-managed urban utilities of Hong Kong, Singapore, Taipei, and Kuala Lumpur pay high salaries, and consequently they tend to attract and retain good managers.

The more efficient utilities also give their managers more financial autonomy and the authority to make prompt and efficient decisions. For example, the Metropolitan Waterworks Authority of Bangkok has the financial autonomy to raise investment funds in the local bond market. Its overall performance is good, and hence the general public subscribes to its bonds. Similarly, the Public Utilities Board of Singapore has considerable autonomy in staffing, finance, and the procurement of goods and services. It also has a clear tariff policy, which is acceptable to the people and politicians as a whole, and thus can be implemented efficiently without any political interference.

The situation in Mexico City is very different, in contrast, especially when compared with Singapore. The head of the water utilities is a political patronage position, and changes with every new mayor, who, until 1997, was appointed directly by the President. The head of the utility is appointed primarily because of his political linkages to the party and the existing political structure, and not for his professional and management expertise. The entire top management structure of the water utility changes with each new mayor, thus preventing the formulation and implementation of any long-term coherent policy and plans. Each new head, even when they are affiliated to the same political party, generally feels major policy changes are warranted. Such regular and radical shifts in policies are not conducive to efficient management on a long-term basis.

Water tariffs became an issue in the first mayoral election in Mexico City during the fall of 1997. The current water tariffs are unlikely to cover even the operating and maintenance costs, let alone the capital costs. The present mayor, who was a candidate at that time, wanted to reduce the prevailing tariffs because he felt the poor could not pay. This is in spite of the fact that the very poor, according to the Mexican Human Rights Commission, currently pay five times more per unit of water than residents in the richest areas, and in addition they have no access to water in their homes, having to buy it from water vendors. The management of urban utilities is unlikely to be efficient until and unless political interference in their management and operation can be significantly reduced. Utilities need to be run by professionally competent managers, within clearly stipulated and transparent guidelines, which could be established by the political process. As long as the utilities adhere to the established guidelines, there should be no political interference with their day-to-day running.

Another issue is the extensive use of public taps in certain major urban centres. This is a good indicator of poor management practices. It is interesting to note that the better-managed water utilities in Asia, such as those in Bangkok, Kuala Lumpur, or Singapore, do not have public taps, because they already have 100 per cent coverage. Public taps often indicate lower levels of service, as well as higher water wastage. In addition, utilities cannot recover revenue from such

Water for urban areas of the developing world

taps, and city authorities are reluctant to subsidize them directly from city taxes.

Environmental and health issues

Water quality and health are major factors that must be considered in efficient urban water and wastewater management.

In the vast majority of urban centres in the developing world, drinking water directly from city supply systems entails considerable health risks. Thus, not surprisingly, people often boil tap water prior to drinking it. In addition, sales of water filters and bottled water have increased, and continue to increase, exponentially in recent years. For example, in India, the sale of bottled water increased by approximately 450 per cent within a short period of six years, 1992 to 1997 (fig. 1.3). This explosion in demand for bottled water is now a common phenomenon in all developing countries, ranging from Brazil to India, and Egypt to Mexico. This demand stems from one and only one factor: the intense fear of consumers, mostly justified, that the water supplied by the public utilities is impure and thus drinking it entails significant health risks.

Ironically, even though consumers have taken to drinking bottled water in ever-increasing quantities, the quality of bottled water often leaves much to be desired. The absence of legal standards and/or the lack of mechanisms for enforcing whatever legal standards do exist

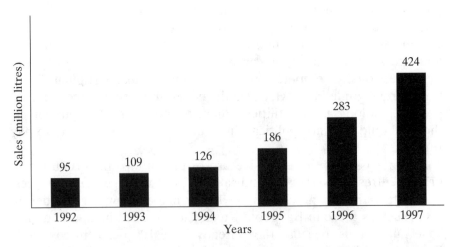

Fig. 1.3 **Use of bottled water in India, 1992–1997 (Source: All India Bottled Water Manufacturers Association)**

mean that quality control is left almost exclusively to the bottlers. Combined with the poor quality control practices of the bottlers, this means that consumers are not getting the safe bottled water for which they are paying.

In India, for example, standards exist for drinking water, but adherence to the standards is purely voluntary. Equally, since there are no specific regulatory requirements for establishing a bottled water plant, anyone who wishes to construct one can do so without indicating the source of the water, the technology used to purify it, and the final quality of the bottled water. The absence of regular, or any, monitoring by the competent authorities further gives the bottlers a free hand. Recently, India's leading weekly news magazine, *India Today* (22 December 1997), carried out an independent analysis of 13 major brands of bottled water. Only three brands conformed to all the specifications for drinking water.

Changes in mind-sets

There need to be some fundamental changes in the mind-sets of the managers of water utilities. Only two will be discussed here.

First, many managers of urban utilities think that a continuous daily supply is not possible because not enough water is available. This contention, however, does not stand up to close scrutiny, as can be seen from the comparisons shown in table 1.4. It should be noted that the statistics used in this table were provided by the utilities of the cities concerned, and thus in all probability present a more optimistic picture than the actual reality.

If per capita consumption in the cities is analysed, table 1.4 indicates that it varies from a high of 265 litres/day in Bangkok to a low of 16 litres/day in Malé. And yet both Bangkok and Malé provide a 24-hour water supply. In contrast, per capita consumption in Delhi is 209 litres/day and in Mumbai 178 litres/day, and still they can provide only 3.5 and 5 hours of water, respectively, each day. This is ostensibly because of a lack of water. Clearly something is fundamentally wrong in such thinking.

The reason for this anomaly is the mind-set of the managers. The basic philosophy of the managers in cities such as Delhi appears to be that, since enough water is not available, it is preferable to provide "some for all" rather than "all for some." In fact, if proper management practices were introduced, it should be possible to provide "water for all" on a 24-hour basis.

Table 1.4 **Water service indicators for selected Asian cities**

	Bangkok	Beijing	Calcutta	Delhi	Hong Kong	Karachi	Kuala Lumpur	Malé	Mumbai	Singapore
Consumption/capita (litres/day)	265	96	202	209	112	157	200	16	178	183
Water availability (hours)	24	24	10	3.5	24	1–4	24	24	5	24
Unaccounted for water (% lost)	38	8	50	26	36	30	36	10	18	6
Coverage (%)	82	100	66	86	100	70	100	100	100	100
Production costs (US$/m^3)	0.173	0.061	0.011	0.037	0.580	0.042	0.131	2.646	0.052	0.309

The residents of cities where an interrupted water supply is the norm are well prepared to cope with the situation. Houses have storage tanks, which are filled whenever water is supplied. For example, in Salt Lake, an upper-class residential area of suburban Calcutta, water is supplied four times per day for a total of around $5\frac{1}{2}$ hours per day. Residents fill up their storage tanks whenever the water supply is resumed, so that houses in reality have a 24-hour uninterrupted water supply. Because water tariffs are very low in many cities, or non-existent (as in Calcutta), it is common to find that the storage tanks are overflowing, which increases water wastage significantly. Current experiences indicate that interrupted services experience more wastage than uninterrupted supplies and per capita water usage is often higher for interrupted supplies compared with a 24-hour continuous supply.

It is evident that cities such as Calcutta, Delhi, or Karachi could introduce 24-hour services with their existing water supplies through radical rethinking of management practices. Technically and economically, it is difficult to see why 24-hour water services cannot be provided in urban areas when the per capita daily supply exceeds 100 litres. After all, cities such as Malé have managed to provide a 24-hour service with a per capita daily supply of only 16 litres.

The provision of a continuous water supply would also change another mind-set of the managers. Past experience indicates that, once the utilities accept the concept of a less than 24-hour water supply, the hours of service provided continue to decline steadily with time, because the managers find this an easier option than making hard decisions, which may irritate their political masters.

The second major mind-set that needs to be changed is the emphasis on new construction rather than reducing the existing water losses from the system. On the basis of information available to me as an adviser to 17 governments, losses of the order of 30–40 per cent are currently the norm rather than the exception. Losses in some cities, such as Oaxaca, Mexico, are of the order of 60 per cent (Arreguín-Cortés, 1994), which is a truly astounding situation.

However, instead of making serious attempts significantly to reduce such losses, the main alternative most often considered is the construction of new water development projects, which would operate under similar inefficiencies. Technically and economically, the reduction of losses is significantly more cost-effective (in fact by several orders of magnitude) than developing new water sources, and yet this alternative has not received the attention it deserves from water

managers and politicians. This irrationality can be explained for the most part by the current mind-sets and interests of three groups of people:
1. *Politicians*, who generally feel that there are more votes in the construction of new projects compared with improving the efficiency of existing systems. Thus, vast sums are allocated for the construction of new projects, but the budgets approved for improving and maintaining system efficiency are starved.
2. *Engineers*, who feel the construction of new projects is exciting, and the operation and maintenance of existing systems should be left to less "bright" people.
3. *Contractors and consulting companies*, who lobby politicians and water managers hard for approval of new projects because enormous funds are needed for the construction of new projects, especially in comparison with the budgets approved for the operation and maintenance of existing systems. They are also one of the largest group of contributors to the political parties.

Unless such mind-sets can be changed, it will be almost impossible to provide safe drinking water and sanitation services to the residents of urban areas in developing countries in the foreseeable future. It should, however, be noted that the total elimination of system losses is not a feasible option. System losses of 7–12 per cent (the situation will vary from one urban area to another owing to a variety of factors) will have to be accepted. Further reductions in losses would mean that the cost of this reduction would start to become progressively higher, and higher than the cost of the water saved. Accordingly, losses of this order have to be accepted. It should, however, be noted that, in England and Wales, not one of the water supply utilities, which are now all in the private sector, has so far managed to reduce system losses to less than 15 per cent.

Concluding remarks

On the basis of the above review, it is evident that the provision of clean water and sanitation to all residents of the urban areas of developing countries will be one of the major challenges of the twenty-first century, the magnitude and the complexity of which no earlier generation has had to face. Regular rhetoric at international forums will not resolve this difficult and complex problem. In the run-up to the twenty-first century, the world really has two choices: to carry on as before with a "business as usual" attitude, which can contribute only incremental changes that would endow future generations with a legacy of inadequate water supply and sewerage services; or to continue in earnest an accelerated effort radically to change the mind-sets of the decision-makers and the water managers so that urban people have access to safe drinking water and sanitation facilities within the next generation. It will not be an easy task to accomplish but, given sufficient political will and proper system management, it could be achieved. One is reminded of William Shakespeare's warning: "Men at some time are masters of their fates: The fault, dear Brutus, is not in our stars, but in ourselves, that we are underlings."

References

Arreguín-Cortés, F. I. 1994. "Efficient Use of Water in Cities and Industry," in *Efficient Water Use,* edited by H. Garduño and F. Arreguín-Cortés, National Water Commission, Mexico, pp. 61–91.

Biswas, Asit K. 1997a. *Water Resources: Environmental Planning, Management and Development,* McGraw-Hill, New York.

Biswas, Asit K. 1997b. "Development of a Framework for Water Quality Monitoring in Mexico," *Water International,* Vol. 22, No. 3, pp. 179–185.

Biswas, Asit K. and Arar, Abdullah. 1988. *Treatment and Reuse of Wastewater,* Butterworths, London.

Bos, E., Vu, M. T., Massiah, E., and Bulatao, R. A. 1994. *World Population Projections, 1994–95,* published for the World Bank by Johns Hopkins University Press, Baltimore, MD.

India Today, 22 December 1997. "Bottled Water: How Safe?" pp. 64–68.

McIntosh, A. C. and Yñiguez, C. E. 1997. *Second Water Utilities Data Book: Asian and Pacific Region,* Asian Development Bank, Manila.

Naranjo Pérez de León, María F. and Biswas, Asit K. 1997. "Water, Wastewater and Environmental Security Problems: A Case Study of Mexico and the Mezquital Valley," *Water International,* Vol. 22, No. 7, pp. 207–214.

Okunlola, P. 1996. "Lagos under Stress," *The Courrier,* November–December, pp. 50–51.

United Nations. 1988. *Population Growth and Policies in Megacities: Karachi,* Population Policy Paper No. 13, Department of International Economic and Social Affairs, United Nations, New York.

——— 1997. "Comprehensive Assessments of the Freshwater Resources of the World," Report of the Secretary-General, Committee on Sustainable Development, 5th Session, E/CN. 17/1997/9, United Nations, New York.

World Bank, 1992. *World Development Report,* Oxford University Press, New York.

2

Water management in Metropolitan Tokyo

Yutaka Takahasi

History of waterworks in Tokyo

Early water supply systems

During the sixteenth and seventeenth centuries, the city of Edo was already equipped with comprehensive water supply systems that did not exist even in Europe. This was one foundation of the prosperity of Edo that has lasted for nearly 300 years. In 1590, Tokugawa Ieyasu ordered Okubo Fujigoro to draw up a master plan for a water supply system, and, based on this master plan, part of the Kanda Canal was completed. In 1654, the Tama River Canal, with a length of 43 km, was completed by using the Tama River, running west of Edo. It became possible to supply water continuously to the centre of the Edo area and its vicinity. These excellent water systems were based on what one might call classical technologies. They depended not on pumps but on the skilful use of gravity flow, and the water was not sterilized. A rough chronological table of waterworks in Tokyo is shown in table 2.1.

Table 2.1 **History of waterworks in Tokyo**

1590	Tokugawa Ieyasu, founder of the Edo Shogunate, commissioned Okubo Fujigoro to carry out a survey and draw up a master plan; partial completion of the Kanda Canal
1654	Completion of the Tamagawa Canal
1898	Yodobashi Purification Plant started operating
1923	Facilities badly damaged in the Great Kanto earthquake
1924	Sakai Purification Plant started operating
1926	Completion of Kanamachi Purification Plant
1934	Completion of Yamaguchi Reservoir
1938	Construction of Ogouchi Dam started
1945	Facilities devastated in World War II
1957	Completion of Ogouchi Dam (Tama River system)
1959	Nagasawa Purification Plant started operating
1960	Higashi-Murayama Purification Plant started operating
1964	Great water shortage in the Tama River system; water distribution cut by max. 50%
1965	Abolition of Yodobashi Purification Plant because of the Shinjuku Suburbanization Plan
1966	Asaka Purification Plant started operating
1967	Completion of Yagisawa Dam (Tone River system)
1968	Completion of Shimokubo Dam (Tone River system)
	Completion of Tone Diversion Weir and Musashi Canal
1970	Ozaku Purification Plant started operating
	Intake stopped from Tamagawa Purification Plant
1971	Completion of Tone Estuary Barrage
1975	Misono Purification Plant started operating
1976	Completion of Kusaki Dam (Tone River system)
1985	Misato Purification Plant started operating
1991	Completion of Naramata Dam (Tone River system)
1992	Completion of 1st Stage of Advanced Water Purification Treatment Facility in Kanamachi Purification Plant
	Completion of Tamagawa Cold Water Countermeasure Facility

Source: Bureau of Waterworks (1994).

Opening of modern waterworks

Along with the opening of the country, the Japanese government imported advanced waterworks technologies developed in Western Europe. Beginning in 1887, modern waterworks were constructed under the supervision of William Palmer, an Englishman, in Yokohama, which had a large settlement of foreigners. The Yodobashi Purification Plant started operating in 1898. At the beginning, it was capable of supplying only 166,800 m^3 of water per day to 80,000

Metropolitan Tokyo

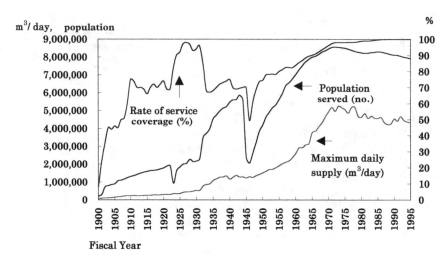

Fig. 2.1 **Developments in Tokyo's water service, 1900–1995 (Source: Bureau of Waterworks, 1994)**

people. However, as the population increased and the potable water service spread, the waterworks kept on growing. Private sector systems and nearby villages and towns were absorbed. Developments in Tokyo's water service are shown in figure 2.1; the growth in the average daily water supply is shown in figure 2.2.

These figures demonstrate the remarkable expansion of the water supply system. The process of expansion was not smooth, however. Some of the difficulties were the Kanto earthquake on 1 September 1923, World War II, and in particular the destruction of waterworks by air raids towards the end of the war, and then slow recovery. Furthermore, serious water shortages in the Tama River in 1940, flood damage in the eastern areas of Tokyo caused by Typhoon Katherine in 1947, and another serious water shortage caused by low precipitation on the Tama River Upper Basin in the summer of 1964 affected the management of the Tokyo water service.

The reconstruction of the water service facilities that had been completely destroyed by the earthquake of 1923 not only employed earthquake-resistant structures, but also involved the construction of new reserve water systems, changing from steam pumps to electric pumps at the Yodobashi Purification Plant, the carrying out of land readjustment for urban renewal, and major modifications to and expansion of water service networks. As a result of the unusual water shortages in the Tama River in 1940, water supply sources were re-

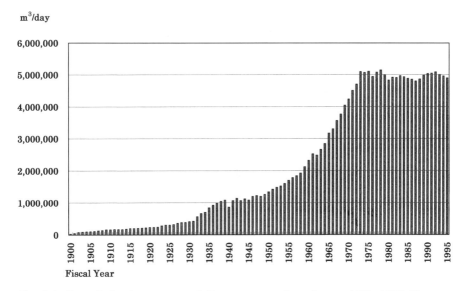

Fig. 2.2 **Growth in the average daily water supply volume, 1900–1995 (Source: Bureau of Waterworks, 1994)**

inforced, new wells were constructed, neighbouring water systems were connected, and emergency measures for water supply were completed.

The Ogouchi Dam project

As a drastic measure to cope with the rapid increase in population, the Tokyo City Waterworks Bureau decided to go ahead with the Ogouchi Dam project. At 149 m, the dam height was, as a dam for the exclusive use of the water service, the highest in the world at the time, and it was an epoch-making undertaking considering that the highest dam in Japan at the time was the Komaki Dam, at 79 m. The water volume of the reservoir (later named Lake Okutama) created by the Ogouchi Dam was approximately 180 million m^3. However, construction did not commence until 1938, because the acquisition of water rights in the lower basin of the river ran into difficulties. Moreover, construction had to be discontinued during World War II because of a shortage of labour and materials.

After the war, construction of the dam recommenced in 1948, and it was completed in 1957. The water supply capability of the Tokyo waterworks immediately expanded. Within a year, the Nagasawa

Metropolitan Tokyo

Purification Plant, intended for bringing water from the Sagami River to Tokyo, was completed. This was built in Kawasaki City in Kanagawa Prefecture, which is next to Tokyo. The ceding of the water supply to Tokyo began in 1959. Also in 1957 the Waterworks Laws were decreed. The promotion of the water service was a significant policy throughout the Japan.

War damage and the increase in water leakage

World War II not only interrupted construction of the Ogouchi Dam, but caused great damage to the waterworks facilities. Air raids became ever more intense over all Japanese cities from 1944 and were heavily concentrated on Tokyo in particular; its waterworks facilities were completely devastated. From August 1945, with the end of the war, reconstruction of the war-damaged waterworks facilities began. Since losses of water from water pipes in Tokyo had been considerable as a result of the war damage and poor maintenance, the repair of water leaks became the main challenge.

Changes in the rate of water leakage since 1915 are shown in table 2.2. The leakage rate had been over 20 per cent since 1930, but in 1945 – at the end of the war – it had rocketed to approximately 80 per

Table 2.2 **Trends in Tokyo's water leakage rate, 1915–1995**

Year	Rate(%)
1915	12.3
1930	21.2
1935	25.8
1945	80.0
1946	68.0
1950	30.0
1955	22.0
1960	22.0
1965	19.2
1970	17.3
1975	16.9
1980	15.6
1985	14.7
1990	12.2
1995	9.9

Source: 1950–1995 – Bureau of Waterworks, *Annual Report on Waterworks in Tokyo*, Tokyo Metropolitan Government.

cent. Though it fell to 68 per cent by 1946, the water supply system was jokingly called the "waterworks colander." Specialists in the Metropolitan Government were employed to find the points of leakage and to deal with the problem. They succeeded in at last bringing the rate down to 22 per cent in 1955. This rate was still quite high. As a result of continued efforts, it fell to 10.6 per cent in 1993. Since the annual distribution of water in 1993 was 1,700 million m^3, a leakage rate of 10.6 per cent meant that the volume of leakage was 170 million m^3 annually, an amount equivalent to the total pondage of the Ogouchi Dam. However, since most of the points of leakage were not in the main pipes, but in the very great number of final distribution pipes, it was not easy either to locate or to deal with the problem; it required time and labour. It took many years to get the rate down to 10 per cent, and ceaseless efforts must continue to be made in the future.

Serious water shortage in 1964

From around 1957, when the Ogouchi Dam was completed, the population of Tokyo started to increase rapidly. This was a time of transition from a period of urban population concentration to a period of high economic growth. The population increase and the rise in living standards resulted in an inevitable increase in the consumption of potable water and the volume of water used for various city activities. For that reason, the annual increase in the water service in Tokyo reached approximately 300,000 m^3.

In 1963/64, during this period of rapid increase in demand for water services, there was little precipitation in the upper basin of the Tama River, the upstream region of the Ogouchi Dam. In particular, precipitation during May and July of 1964, the rainy season, was extremely low. At this time, the sources of water for Tokyo consisted of the Tama River system, including the Ogouchi Dam (about 60 per cent), the Edo River (about 20 per cent), the Sagami River (about 10 per cent), and underground water and other sources (about 10 per cent). Dams along the upper basin of the Tone River, such as the Yagisawa Dam, were under construction, and others were still in the planning stage, so water from these areas was not yet supplying the water service system to serve the citizens of Tokyo. The lack of precipitation in the upper basin of the Tama River system lasted from June until 20 August 1964, and the pondage of Ogouchi Dam, which was 180 million m^3 at its maximum, fell to as little as 2 million m^3.

This was a decisive blow to Tokyo's water supply. The Tokyo Waterworks Bureau had to impose severe restrictions on the use of water. Areas in the highlands of Tokyo went whole days with no water. Water tankers of the Self Defence Forces were mobilized every day, and citizens had to wait in line with buckets for their share.

The Olympic Games, held for the first time in Asia, were to open on 10 October 1964. A construction rush had been under way in Tokyo for hotels, metropolitan expressways, the Tokaido Shinkansen, and subways. Construction sites and newly built hotels were also suffering from the shortage of water.

Because the dams along the upper basin of the Tone River were not yet completed, the construction schedule for the Musashi Canal (a man-made channel connecting the Tone River and the Ara River) was accelerated. As part of the total water system carrying water to Tokyo, it succeeded in achieving a temporary supply of water by 20 August. Luckily, toward the end of August, normal precipitation for the season fell in the upper Tama River basin, and restrictions on water use were gradually lifted. The Olympic Games went ahead without any water-related fears.

Development of water resources in the upper Tone River basin

Development of the water resources of the upper basin of the Tone River system to serve Tokyo had been proposed to Tokyo City Council in 1926. Concrete discussions in the Council started in 1936, but it was not until after World War II that the actual plans were approved for execution. In March 1963, a plan to bring water from the Tone River to Tokyo was decreed by the Cabinet as the "Water Utilization Plan of Tone River Systems." Based on the plan, the development of water sources has become a part of the National Water Resource Development Plan, and many of its projects have been executed by the Water Resources Development Public Corporation that was established in 1962. The Tokyo Metropolitan Government was to take partial financial responsibility for the costs of the water supply, including industrial water, by way of multi-purpose dams. The Akigase Intake Weir and Asaka Canal had been built as emergency measures during the "Tokyo Water Famine" of 1964 as described earlier. The Musashi Canal was also constructed by the Public Corporation. Thus it became possible to get water from the Tone River systems when there was some spare volume, until the completion of dams in the upper basin of the river. Water from the upper Tone River basin was planned to flow through the Musashi Canal, via Tone Oseki (Tone Grand Diversion Weir), to the Ara River, with purification occurring at the Asaka Purification Plant, and sent from there to Tokyo by way of water pipes. The Asaka Purification Plant was completed in 1966. (Prior to that, in 1965, the Yodobashi Purification Plant, which had played an important role as the largest plant in Tokyo, ceased to exist. The site occupied by the plant was taken over for the development of the Shinjuku Suburbanization Plan, and became a town of high-rise buildings, such as the Tokyo Metropolitan Government Centre, hotels, and offices.)

The volume of water supplied to Tokyo increased by 1,200,000 m^3/day after construction of the Asaka Purification Plant (900,000 m^3/day) and the Higashi-Murayama Purification Plant (300,000 m^3/day). All planned construction was completed by 1968, including the Yagisawa Dam in the upper basin of the Tone River in August 1967 and the Shimokubo Dam in the basin of the Kanna River, a branch of the Tone River, in November 1968. As a result, the volume of water supplied to Tokyo increased dramatically. Furthermore, the Tone River water resource development projects were completed one by one, and after 1965 the water supply operation expanded to serve not only the urban areas of Tokyo but also Tama districts. Construction of the Tone Estuary Barrage was completed in 1971 and dams in the basin of the Watarase River, the Kinu River, etc. and expansion of the Asaka Purification Plant were undertaken in the 1960s and 1970s.

Since the 1970s, however, it has become increasingly difficult to get agreement on dam-site areas. Difficulties in balancing water demands in the future have been anticipated. The Tokyo Waterworks Bureau made a public announcement in 1973 on "Statements Concerning Water Conservation" and, for the first time, appealed to citizens regarding the need for control over water demand and the saving of water. At about the same time, the state government began to propose a "water conservation conscious society." Furthermore, the 1973 "Act of Special Measures for Reservoir Areas," a measure to cope with the difficult situation of upstream reservoir areas, was passed by the Diet. Great progress has been made, by making a differentiation from money compensation-type measures. This was one of the turning points in the history of dam construction policy.

Criticisms about dam construction were originally based on the shortcomings of the measures for the reservoir areas of the upstream basin. Eventually, the effect of dam construction upon the environment has begun to be taken into account, with the cost of environmental measures being added on. The cost of construction has thus risen considerably. Although the Yanba Dam and other dams along the Agatsuma River (a branch of the Tone River) are already on the government's construction list, potential dam sites are in general decreasing and it is gradually becoming difficult to secure future water resources for Tokyo by means only of dams.

Tokyo's waterworks, having experienced a century of many complications, have fulfilled their mission well. Now there are new problems: environmental problems such as pollution of water at intake points on rivers, etc., further upgrading of service to inhabitants, and earthquake measures.

The present situation

General view

The waterworks of Tokyo have become a gigantic and complicated system supporting a modern megalopolis with a population of 12 million. In this section, present-day conditions will be outlined and comparisons made with other cities in Japan and with some of the major cities of the world.

The water resources of the present Tokyo waterworks are based on three major river systems, namely the Tone River system (80.2 per cent), the Tama River system (16.7 per cent), and the Sagami River (2.9 per cent), and underground water (the remaining 0.2 per cent). Classification of these, in relation to the purification plants, is shown in table 2.3. Most of the dams and purification plants relating to the Tone River system development were completed after the latter half of the 1960s, and these supply most of the water requirements of the citizens of Tokyo.

The dams in the Tama River system and Tone River system, used for supplying water, are listed in table 2.4. In addition to the Ogouchi Dam in the Tama River system, three reservoirs were constructed in the 1920s and 1930s; these store water taken from Intake Weir of the Tama River system for a short period of time, and send it to the Higashi-Murayama and Sakai purifying plants. In 1996, the main distribution pipes were 2,009 km in length, and the small pipes were 19,887 km in length, giving a total length of 21,896 km (see table 2.5).

One of the important tasks of today's purification plants is to pro-

Table 2.3 **Purification plants for Tokyo's water supply**

Water resources	Purification plant	Capacity (10³m³/day)	Contribution (%) Plant	Contribution (%) System	Treatment method	Completion
Tone River system	Kanamachi	1,600.0	23.0		Rapid Sand Filtration	1926
				80.2	Partial Advanced Water Treatment	
	Misato	1,100.0	15.8		Rapid Sand Filtration	1985
	Asaka	1,700.0	24.4		Rapid Sand Filtration	1966
	Misono	300.0	4.3		Rapid Sand Filtration	1975
	Higashi-Murayama	880.0	18.2		Rapid Sand Filtration	1960
Tama River system	Ozaku	280.0	4.0		Rapid Sand Filtration	1969
	Sakai	310.0	4.5	16.7	Slow Sand Filtration	1923
	Kinuta-kami	11.5	1.7		Slow Sand Filtration	1928
	Kinuta-shimo	70.0	1.0		Slow Sand Filtration	1922
	Tamagawa[a]	(152.5)	—		Rapid Sand Filtration	1917
					Slow Sand Filtration	
Sagami River system	Nagasawa	200.0	2.9	2.9	Rapid Sand Filtration	1959
Underground water	Suginami	15.0	0.2	0.2	Chlorine feeding	1932
Total		6,956.5	100.0	100.0		

Source: Bureau of Waterworks (1994).
a. Production at the Tamagawa Purification Plant has been halted because of pollution of the Tama River.

Table 2.4 **Dams in the Tama River and Tone River systems**

Name	Effective capacity ($10^3 \, m^3$)	Catchment (km^2)	Dam Type	Height (m)	Length (m)	Completion
Murayama-kami Reservoir	2,983	1.3	Earth dam with impervious wall	24	318	1924
Murayama-shimo Reservoir	11,843	2.0	Earth dam with impervious wall	33	587	1927
Yamaguchi Reservoir	19,528	7.2	Earth dam with impervious wall	35	691	1934
Ogouchi Reservoir	185,400	262.9	Non-overflow straight concrete dam	149	353	1957
Fujiwara Dam	35,890	233.6	Gravity system	95	230	1957
Aimata Dam	20,000	110.8	Gravity system	67	80	1959
Sonohara Dam	14,140	439.9	Gravity system	77	128	1965
Yagisawa Dam	175,800	167.4	Arch system	131	352	1967
Shimokubo Dam	120,000	322.9	Gravity system	129	303	1968
Kusaki Dam	50,500	254.0	Gravity system	140	405	1976
Watarase Reservoir	26,400	–	Pit-type reservoir	–	–	1989
Naramata Dam	85,000	95.4	Rock-fill	158	520	1991

Source: Bureau of Waterworks (1994).

Table 2.5 **Features of Tokyo's water service, 1984/5–1993/4**

	1984/5	1987/8	1990/1	1993/4
Population served (10^3)	10,919	11,019	10,973	10,928
Rate of service coverage (%)	99.7	99.9	100	100
Distribution pipe length (km)	19,280	20,164	20,884	21,484
Facility capacity ($10^3 m^3$/day)	6,079	6,629	6,629	6,959
Gross annual supply ($10^6 m^3$)	1,743	1,696	1,773	1,763
Maximum daily supply ($10^3 m^3$)	5,777	5,485	5,955	5,737
Average daily supply ($10^3 m^3$)	4,775	4,634	4,858	4,830

Source: Bureau of Waterworks (1994).

duce water that is safe and tastes good, no matter how polluted the original supply. In particular, the Kanamachi Purification Plant, where the water comes from the Edo River, was no longer able to supply safe and palatable water with the use of the conventional rapid filtration system. A mouldy smell had been noticed in the water since around 1972, and filtering the water through powdered activated carbon had been tried without satisfactory results. Hence, an advanced water purification treatment system combining ozone, biochemical, and activated carbon treatments has been employed. This system is capable of treating 520,000 m^3 (approximately one-third of the plant's total daily capacity of 1,600,000 m^3). In order to cope with a similar problem that started in 1994 at the Misato Purification Plant, where the water is also taken from the Edo River, the same advanced water purification treatment system is under construction to exclude ammonia-based nitrogen gas, which is the cause of the smell of mould and bleach.

Since the pollution of river water has developed, advanced water purification technologies have been sought, and the cost of water purification has risen. Similar problems can be observed in Osaka's waterworks, where pollution is becoming serious in its water sources: Lake Biwa and the Yodo River.

In order to meet changes in the social environment appropriately and to respond promptly to diversified needs, new technology is developing. Currently, the most important areas are: the development of purification technology; improvement of the transmission and distribution system; improvement of the direct supply system; improvement of leakage prevention technology; and the effective utilization of resources and energy.

Industrial water

Industrial water started to be supplied in Tokyo at the time of high economic growth when the demand for industrial water was rapidly increasing. It started in the Kotoku District in 1964 and the Johoku District in 1971. The problem of land subsidence due to pumping an excessive amount of underground water had been serious in these eastern areas of Tokyo. In order to stop the pumping of underground water, industrial waterworks were constructed as a substitute. These measures, in place since 1975, have been seen as one of the main reasons for the cessation of subsidence in the eastern areas of Tokyo.

However, since demand for industrial water has been decreasing since 1974, owing to the relocation of factory sites, water conservation policies, etc., some of the water has been converted to use in incineration plants, for car washing, and for flushing toilets. Even so, the volume of industrial water use is still at excessive levels, and the accounts of the industrial water industry have continued to be in the red. Along with the problem of ageing facilities, a complete restructuring of the operation must be sought.

Utilization of treated sewage

The increase in the volume of sewage is proportional to the increase in water demand. The rate of coverage of the sewerage system in Japan passed 50 per cent in 1995. Japan is still a developing country as far as sewage is concerned. However, if one considers just the Ward Areas in Tokyo, the rate is now 100 per cent.

Treated sewage is beginning to be utilized for a variety of purposes, including toilet flushing (but not for drinking water). Treated sewage has also been utilized to cope with the exhaustion, even in normal conditions, of the Tamagawa Canal and the Nobidome Canal, which were constructed in the Edo period. A daily volume of about 43,000 m^3 has been transferred from the sewage treatment plant in the upper basin of the Tama River into the Nobidome Canal, starting in 1985, and into the Tamagawa Canal, from 1986. Since 1984, a maximum daily volume of 8,000 m^3 of treated sewage from the Ochiai Sewage Treatment Plant has been supplied to the Shinjuku Subcentre Area, where the Yodobashi Purification Plant used to be located.

Furthermore, since 1995, treated sewage has been utilized to supply water, at a rate of 1 m³/sec, to the Shibuya River, the Meguro River, and the Nomi River, where the volume of flow in normal conditions has shown a marked decrease.

The biggest cause of the decrease in flow in these streams, at normal times, has been the spread of the sewerage system. The sewerage system in Tokyo is designed to cope with heavy rain of 50 mm/hour. Consequently, most rainwater is drained through the sewerage system, and drainage by small streams has become unnecessary, even at times of heavy rain. In other words, most of the surface flow has been transformed into underground flows.

It is ironic that the treated sewage from sewage treatment plants has been put back into these streams in order to maintain their environmental balance, when discharges have decreased markedly because of the spread of the sewerage system. In some cities in Europe, small streams have been revived by discontinuing the use of sewerage systems. Sewerage systems have been a symbol of civilization, but this is no longer the case. One's view of twenty-first-century civilization is now affected by the flow of treated sewage into streams that have lost much of their original water volume.

Towards a "water conservation conscious city"

In 1973, the "oil shock" created economic confusion throughout the world. Japan, having been seriously affected, has since then employed energy conservation as a state policy.

In January 1973, in order to balance water demand and also to control it, the Tokyo Waterworks Bureau publicly announced its "Policy to Control Water Demand," which was a forerunner for cities across Japan. At that time, despite the increase in water demand in Tokyo, water resource developments were not progressing according to plan, owing to popular movements against dam construction. Future water supply shortages were therefore anticipated.

For 80 years, with the constant increase in public water demand, the bureau had been planning and executing water resources development projects. This new policy was a great turning point in Tokyo's water demand planning. Furthermore, in Japan the supply of drinking water is a financially independent business within each city. Considering the fact that controlling demand meant less revenue, this was a drastic change of policy in the Waterworks Bureau.

At about the same time, a "water conservation conscious society" was proposed by the water administration of the state government. Such a policy was becoming popular all over Japan under the influence of cities such as Tokyo where water consumption was high.

In 1987, learning a lesson from the water shortages of that year, a Round-table Committee for Creating a Water Conservation-Conscious Society was formed within the Waterworks Bureau. The committee's report states the need to make urban society aware of the need for water conservation through the reinforcement of conventional water conservation systems and the philosophy of water recycling. Since then, the Waterworks Bureau has been actively promoting public relations activities to develop awareness of water conservation among citizens in their daily lives. It has also been requesting manufacturers to develop fixtures such as faucets, toilets, and laundry machines that conserve water.

Concerning the promotion of efficiency in water use, since 1984 there have been individual building recycling, district recycling, and large area recycling. Treated sewage and industrial water have begun to be utilized as their water resources.

As part of its leakage preventive measures, the Bureau makes it a rule to carry out repair work on the day that a surface leak is found. Where the leak is underground, the potential leakage volume is assessed by the minimum flow measurement method, and leaks are located with electronic leak detectors, correlation-type leak detectors, etc. (all performed at night-time). As a result, the leakage rate was reduced to 16.1 per cent in 1977 and to below 10 per cent in 1995 (as shown in table 2.2 above). The target is to bring the rate down to 7 per cent by the beginning of the twenty-first century. To prevent leaks, ductile metal and stainless steel are being used for water distribution pipes.

Tokyo's water service compared with other cities in Japan and the world

Tokyo, the capital of Japan, is the biggest city in Japan. It covers an area of 2,183 km^2, which is 0.6 per cent of the total national land area. Its population of a little short of 12 million is 9.5 per cent of the total national population. Its population density of 5,500 persons/km^2 is about 17 times the national average.

Its water service population, waterworks capacity, and volume of water distributed are compared with other major Japanese cities in table 2.6. The maximum daily consumption of water per person in

Table 2.6 **Features of the water service in Japan's main cities, 1994**

City	Population served (10^3)	Average daily supply per person (litres)	Maximum daily supply per per on (litres)	Water supply facility capacity ($10^3 m^3$/day)	Household rates (yen/$10^3 m^3$)	Length of distribution pipes (km)
Sapporo	1,706	314	381	785	1,194	4,826
Sendai	922	382	460	463	1,266	2,824
Kawasaki	1,193	437	509	1,026	587	2,135
Yokohama	3,310	399	486	1,780	587	8,406
Nagoya	2,146	386	478	1,424	570	4,990
Kyoto	1,426	485	619	980	700	3,598
Osaka	2,603	580	729	2,430	772	4,993
Kobe	1,504	415	503	833	762	4,172
Hiroshima	1,091	382	490	644	576	3,627
Kita-Kyushu	1,018	360	441	710	751	3,544
Fukuoka	1,214	296	386	705	927	3,372
Tokyo	10,928	430	513	6,960	791	21,484

Source: Bureau of Waterworks (1994).

Metropolitan Tokyo

Table 2.7 **Features of the water service in various cities of the world**

City	Population served (10^3)	Length of distribution pipes (km)	Length of distribution pipes per 10^3 people served (km)	Maximum daily supply per person (litres)
Bangkok	4,800	8,086	1.7	479
Singapore	2,558	3,905	1.5	250
Cape Town	2,200	3,094	1.4	426
Rome	2,830	4,810	1.7	636
Vienna	1,470	2,950	2.0	393
Geneva	304	911	3.0	829
Rotterdam	1,100	2,700	2.5	–
Detroit	3,469	5,517	1.6	1,764
Tokyo	10,928	21,484	2.0	513

Source: Bureau of Waterworks (1994).

some of the major cities of the world is compared in table 2.7. Compared with Tokyo, the water supply volume of Detroit in the United States is very large, whereas in other cities consumption is lower. This reflects differences in attitudes and habits toward the use of water in each city. Nevertheless, in the face of severe environmental problems, every city should make efforts to save water.

The future

New targets for water supply works

Ever since the establishment of modern waterworks in Tokyo a hundred years ago, efforts have been made to secure water resources and to maintain facilities. These efforts have brought results at last, but it is no longer a question of just securing the necessary volume of water. A series of new problems has surfaced. To cope with these problems, in 1997 the Tokyo Waterworks Bureau proposed seven significant targets for the next quarter-century.

Waterworks that are immune to shortages
In recent years, there has been no shortage as serious as that of summer 1964. However, every few years, restrictions on water use have been imposed in periods of low rainfall. Because water is used in many diverse fields, the effect of water restrictions on citizens' lives and activities is quite serious.

Metropolitan Tokyo

The goal for Japan's waterworks is a stable supply of water even in times of great shortage that occur once every 10 years, but Tokyo's waterworks have not yet reached that level. The waterworks of San Francisco and New York in the United States have been designed to withstand the greatest historical shortages, and in London they are designed to withstand the shortage that occurs once in 50 years. The reservoirs on the Tone River and the Tama River for Tokyo's waterworks hold about 30 m^3 per person. In comparison, the pondage per person is 520 m^3 for San Francisco and 280 m^3 for New York. The pondage is as low as 90 m^3 per person for Paris and 35 m^3 for London, but the Seine River, as the water resource for Paris, and the Thames River, as the water resource for London, experience little fluctuation in discharges, which has made a stable supply possible. Since the pondage per person for the Tokyo waterworks is quite small, it can be said that the safety margin of the water supply is not high. It is therefore important to establish a waterworks system immune from shortage.

Waterworks that can provide a constant water supply
Many disasters related to water quality have occurred in recent years – there were 299 cases in 1995. About 60 per cent of cases of water pollution are due to oil. In order to cope with the problem, channels for emergency communications and information collection have been established by the communications network through conferences held by related administrative organizations.

Disasters related to waterworks facilities, in particular purification plants, include pollution of water sources with toxic oils, ageing of the facilities, and electricity failure caused by lightning strike or snow fall. Disasters related to water pipes involve traffic vibration, ageing of pipes, leaks caused by soil corrosion, and damage from construction works such as road repairs or gas pipe works.

Even in times of disasters related to water quality or facilities, systems to ensure a constant water supply must be established.

Preparation for a great earthquake
The Hanshin-Awaji earthquake of January 1995 should be acknowledged as a precedent for epicentral earthquakes in Kanto urban areas. It is clear that Tokyo's dilapidated water pipes would be devastated if hit by such an earthquake. Since the Hanshin earthquakes, the Tokyo Waterworks Bureau has been proceeding with the reinforcement of reservoir, intake, purification, transmission, and distri-

bution facilities against seismic shocks. In order to secure potable water in the event of an emergency, it is planned to locate water storage bases every 2 km. For this purpose, existing purification plants and water stations will be used as water storage bases. For areas that are more than 2 km from these purification plants and water stations, emergency water tanks are already in place; for example, there are 45 tanks in the Ward Areas and 7 tanks in Tama District. Each tank contains 1,500 m^3 of water and is placed in a park that has been designated as a refuge. Within the Tokyo area, there are 169 emergency water supply points, and the total potable water volume constantly stored is 910,000 m^3. This is equivalent to the consumption of Tokyo's 12 million citizens for three weeks, allowing 3 litres of water per person per day.

Taking into account the worst possible scenario at the time of an earthquake disaster, ways to secure potable water and water to extinguish fires must be sought.

Maintaining water quality
As regards Tokyo's future water management, measures to maintain the quality of the water must be seriously considered. The advanced water treatment started at the Kanamachi Purification Plant should not be considered as a temporary measure to cope with the contamination of water resources. It should be seen as the forerunner of measures at a time when many new and dangerous chemical substances are being developed.

Providing an impartial and efficient supply
In normal times, or even in times of disasters or shortages, the purpose of the water service is to provide an impartial and efficient water supply for users. Means of establishing such a system must be sought.

Waterworks that consider the environment
Waterworks must be designed to take into account energy saving, the efficient use of energy, and the recycling of resources at all stages such as purification treatment and the supply and operation of water services.

Waterworks that are familiar to users
In order to realize a peaceful life for the users, information collection from users and a give-and-take two-way information system must be

encouraged. It is important for waterworks to become familiar and intelligible to its users.

Future water resources policies

Conventionally, the concept of water resource developments was limited to the production of new water resources through river developments, including the construction of dams and estuary barrages. For now, and for some time to come, these conventional river developments will continue to be the main technologies. But the demand for and supply of water should be brought into balance by combining various developing technologies and not by depending on river developments alone.

Considering that water resources are circulating resources, water at every phase of circulation must be seen as a resource. In other words, water sources should not be limited to the water from rivers, lakes, and ponds and underground water, but should include all forms, from rainwater to treated sewage.

The utilization of treated sewage for building use and environmental use, which has already begun on a small scale in Tokyo, is a significant step in the long-term vision of future water resource policies. Though there are many problems with using treated sewage, such as cost, administrative matters, and the creation of laws, its utilization must become the most important task of the twenty-first century. There are many ways to attain this goal, such as sending treated sewage back to the upper basins of rivers, or sending it through underground pipes to wherever it is needed, as is being done in some areas already.

Using treated sewage has many advantages, including the facts that it enables the increase in water demand to be met, and that the production of treated sewage is carried out close to the place of water consumption. The utilization of treated sewage is a good way of increasing the rate of water recycling and thus contributing to the ultimate goal of efficient water utilization.

As far as the desalination of seawater is concerned, the costs of constructing and operating desalination plants are quite high, and energy consumption at plants is very high. Furthermore, for Tokyo it would be necessary to provide extremely long pipes into Tokyo Bay to get clean seawater. All these problems make the idea unfeasible for some time to come.

The development of water resources through conventional river projects is reaching its limits for large cities such as Tokyo. Dam sites are getting further away from consumption areas, and the effects of dam developments upon the natural and social environment have to be stringently watched. Consequently, the cost of measures to deal with environmental problems has made the cost of dam construction high.

Promoting awareness of water conservation among users will be an important part of water resources policies in the future. Since the development of water resources has become expensive and difficult, controlling the rise in water demand is vital. To this end, water users must be made aware of the fact that water is an invaluable resource and that it must be used sparingly. The use of water-conserving appliances must be expanded, and public relations activities must be reinforced. Such efforts should not be limited to water resource areas, but should become an essential measure in coping with the deterioration of the global environment.

Water circulation and urban civilization

Rapid urbanization and changes in water circulation

Urbanization affects water circulation. The spread of sewerage systems, as described previously, caused a deterioration in river and canal environments as a result of water flow loss. Other changes in water circulation are caused by the paving of roads and the conversion of farmland into housing.

The urbanization of Tokyo started in the 1950s. Tokyo's population increased rapidly from the latter part of the 1950s into the 1960s, and is now some 12 million, warranting the name "mega-city." Urbanization at such a pace has changed the water circulation of Tokyo drastically, and has become the cause of new urban flood hazards. The Kano River typhoon, on 26 September 1958, produced the highest recorded rainfall per day (392 mm) since 1875 in Tokyo, and caused great flood damage in the newly developed housing areas of the western part of Tokyo. Since then, damage caused by rainfall has increased in parallel with new housing developments. Changes in water circulation during heavy rains are the main cause.

The urbanization of Tokyo has been accompanied by the populace's desire for a higher standard of living, resulting in a heavy burden on rivers and water circulation. These burdens have involved the control of water, the utilization of water, the environment, and the landscape. Embankments became taller as a result of river improvement works undertaken to protect against water hazards, thus spoiling views along rivers and streams. The construction of highways to provide easy access to the areas alongside these rivers has also spoiled riverside scenery.

In today's Tokyo, projects that aim to restore the rivers and water circulation of the city are finally under way. They include advanced water treatment systems in purification plants, the discharge of treated sewage into rivers and streams, the utilization of treated sewage in high-rise buildings, the encouragement of urban renewal work, and river improvement work in the development of the Super Embankment along the Sumida River – a river that is emblematic of Tokyo.

The philosophy of recycling

Urban developments that contribute to the convenience of urban life and economic efficiency have altered the nature of water circulation in Tokyo. As a result, the populace has been troubled by new types of flood hazard since 1985, a decrease in water bodies, a decrease in the ability to control the temperature in the city, and the heat island phenomenon, which has become acute in recent years. Tokyo waterworks have eagerly sought to meet the increased demand for water and have developed water resources by means of dam construction. New technologies, based on visions worthy of the twenty-first century and not limited to conventional planning ideas, have been sought for the water management of the future.

This could be called the materialization of the philosophy of recycling. The characteristics of water as a natural resource are intrinsic in the meaning of the recycling of resources. The utilization of treated sewage in buildings and to recharge rivers since the latter half of the 1980s in Tokyo, no matter how limited, should be recognized as the forerunner of water recycling measures from the point of view of the history of technologies.

New technologies must be developed to cope with the utilization of treated sewage, which is expected to grow in volume in Tokyo in the future. In order to achieve this goal, it is important to offer water of acceptable quality at low cost. The administration that produces treated sewage and the administrative bodies for waterworks, rivers, and streams, and the environments expected to use the treated sewage are related to each other. In order to realize the philosophy of water recycling, an all-around administration is a must. In order to achieve this academically, the development of interdisciplinary fields of studies and cooperation is required. At the moment, treated sewage is sent underground from treatment plants to buildings, rivers, and water channels. For the future, however, studies are already under way on numerous technologies for sending it back to the upper basins of rivers and streams supplying purification plants.

In this context, the utilization of treated sewage must be recognized as one part of water resource development. Water resource development for big cities in the future should be a combination of dams, the utilization of treated sewage, the use of rainwater, as is being done in Sumida District, Tokyo, the conversion of existing water rights, etc. The designers and the executors of development plans must recognize that projects that do not take recycling into account affect the natural circulation of water in that area. If this concept is not accepted, the philosophy of recycling will not be able to be applied to Tokyo as a basic element in city planning.

The concept of recycling should not be limited to the waterworks of Tokyo; it should be applied to the water management of any future megalopolis and especially to the future planning of water-related infrastructure. As water-related projects become bigger and more complicated, the concept will expand its influence beyond borders. Hence, the concept is undoubtedly the key to the global problems of water and the environment.

Acknowledgements

I am grateful to Mr. Isao Tahara, an engineer with the Waterworks Bureau of the Tokyo Metropolitan Government, for collecting the data on waterworks in Tokyo, and to Mr. E. Marui and Mr. S. Konda, engineers with Construction Project Consultants, Inc., for their English language assistance in preparing this article.

Bibliography

Bureau of Waterworks. 1994. *'94 Waterworks in Tokyo*, Tokyo Metropolitan Government.
―――― 1996. *Annual Report on Waterworks in Tokyo*, Tokyo Metropolitan Government (in Japanese).
Japan Waterworks Association. 1996. *Statistics on Waterworks in Japan* (in Japanese).

3

Water quality management issues in the Kansai Metropolitan Region

Masahisa Nakamura

Introduction

The Metropolitan Region of Kansai, which encompasses the cities of Kyoto, Osaka, and Kobe and surrounding municipalities in the Lake Biwa–Yodo River–Osaka Bay water system (see fig. 3.1),[1] has a population of some 18 million, of whom 14 million are served by the water of Lake Biwa (see table 3.1). The region is also characterized by major industrial developments, particularly along the Osaka Bay and Seto Inland Sea strip, as well as by extensive paddy agriculture in the Lake Biwa lowland areas. A complex web of water supply and wastewater networks serves and supports the region and its high level of municipal, industrial, and agricultural activities.

The Kansai Metropolitan Region has historically been dependent on Lake Biwa and the Yodo River for its water resource needs. The Biwa–Yodo system, therefore, had to be gradually transformed over the past century or so from a natural water system to a managed water system, through the installation of flood control facilities and the development of water resource management and control systems. The latest and the greatest of the sequence of engineering projects for water management is called the Lake Biwa Comprehensive Development Project (LBCDP), which was completed in March 1997

Kansai Metropolitan Region

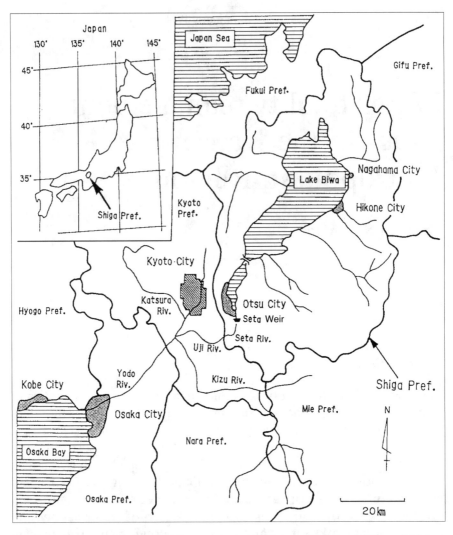

Fig. 3.1 **The Lake Biwa–Yodo River–Osaka Bay area and part of Kansai Metropolitan Region**

after 25 years. Thanks to the construction of a new weir and of levees around the lake, Lake Biwa water can now be discharged into the Yodo River in much greater quantities and with less variability to satisfy urban, industrial, and agricultural needs around and downstream of the lake at times of severe drought. Flooding problems can also be dealt with more easily now by resorting to newly installed pumping facilities and control gates for all of the rivers and irrigation

Table 3.1 **Population served by Lake Biwa water, 1994**

Prefecture	Population within jurisdiction	Population served by Lake Biwa water	Dependency on Lake Biwa water (%)
Shiga	1,289,277	1,012,185	79
Kyoto	2,602,351	1,789,509	69
Osaka	8,719,584	8,525,052	98
Hyogo	5,466,316	2,612,801	48
Total	18,077,528	13,939,547	77

channels that flow into Lake Biwa (Nakamura and Akiyama, 1991; Nakamura, 1995).

The Kansai Region, however, has still to deal with many unresolved problems of water management, particularly with respect to water quality. The improved water resource systems and infrastructures around Lake Biwa will promote further development of the watershed that will bring about environmental issues of greater magnitude and of a more complex nature. Water quality issues, which are by themselves inherently quite difficult to deal with owing to their growing complexity, are now intertwined with quantity issues, making the management of the Biwa–Yodo–Osaka Bay system extremely challenging both for the upstream and for the downstream regions.

This chapter will attempt to highlight some unique features of the water system in the region and to describe briefly its evolution as well as the implications for the current physical and institutional structures. It will also attempt to address the issues facing individual subregions and municipalities as well as the issues facing the entire water system within the broad context of water quality management, e.g. upstream–downstream relationships, the upgrading of the wastewater management system, the water quality of Osaka Bay, and control of Lake Biwa eutrophication, all with reference to such emerging concerns as sustainable water use, integrated watershed management, and the attainment of a sound ecosystem.

The Kansai Metropolitan Region

The Kansai Metropolitan Region includes such major cities as Osaka (with a population of 2.64 million), Kyoto (1.50 million), and Kobe (1.41 million) and many surrounding municipalities, including some medium-sized cities such as Otsu, Nara, and Wakayama. It boasts the third-largest industrial output in Japan, amounting to 40.5 trillion yen (in 1990) or about 12.4 per cent of the total domestic product.

Throughout history, Osaka has been known as a merchant city and also as a place where many new businesses sprout. Although much of the city was destroyed during the Second World War, it made a rapid recovery and achieved remarkable industrialization in post-war decades. The industrial complexes extending south as well as west along the Osaka Bay coastline formed what is called the Hanshin Industrial Complex. At the same time, a large number of small-scale manufacturing industries have sprung up in the mixed residential and commercial districts within the city as well as in the suburban regions, resulting in uncontrolled urban and semi-urban growth into the surrounding regions. Such patterns of development are said to have caused the disappearance of the once extensive canal network around the Bay which was used for shipping commodities up through the Yodo River, the reason for Osaka having once been called the Venice of the Orient.

Kyoto, the ancient capital city, has been endowed with many cultural and scenic assets throughout history. It has the highest concentration of temples and shrines in Japan, which are major tourist attractions. The city, however, had been constrained by its water resources, and industrialization of the city in the late nineteenth century lagged far behind that of downstream Osaka. Prompted by this observation, the governor at the time initiated a study on the construction of a canal to link the city with Lake Biwa, which was completed in 1891. Kyoto began to regain its economic strength and was able to develop various manufacturing industries other than the traditional kimono textile and *sake* (rice wine) industries. Since the construction of the canal, Kyoto has never experienced serious water shortages.

Kobe, a port city some 30 km or so west of Osaka, has thrived because of its role as a major port for international trade. However, it is extremely constrained in terms of space because it is situated on a strip of coastal land backed by a range of plateau land. The city has undergone tremendous transformation in recent decades through the reclamation of its coastal shallows. Some major steel producers and ship builders once dominated the industrial structure of the area, but there is now a fairly significant number of small to medium manufacturers of low-cost household goods. Yodo River water transported from Osaka helped to sustain the population and industry. An earthquake in January 1995, however, devastated the city and much of its infrastructure, including its industrial facilities. The city is now mobilizing every available resource to recover from the damage and to reconstruct the whole of the metropolitan system, including water and wastewater facilities.

Around and between these great cities lie many small municipalities, whose boundaries are hardly noticeable owing to widespread urban development. Many industries that used to operate within the inner metropolitan districts have been relocated steadily since the war to the outer districts, resulting in the creation of these new municipalities with growing populations. In the Metropolitan Region of Kansai, the management of water, with respect to both quantity and quality, has become extremely important as well as intricate. Each of the three main metropolises has had its own unique problems to overcome and has developed its own unique system of water services. The wastewater from these urban developments has to be collected, treated, and discharged. These discharges, together with urban runoff after rain, find their way into tributary water channels prior to reaching the main watercourse of the Yodo River. In a short stretch of less than 30 km, the discharges and water intakes have to compete before the river water finally reaches Osaka Bay, about which we learn in some detail in the next section.

The water resources in the region

The Lake Biwa–Yodo River water system

The Kansai Metropolitan Region lies in the central part of the area known as the Kinki District, where the natural setting differs widely between the north, which faces the Sea of Japan, and the south, which faces the Pacific Ocean. The weather is generally much warmer and drier in the south than in the north, where annual precipitation may exceed 3,000 mm. In the central lowland areas, Lake Biwa collects runoff from the precipitation on the surrounding mountain ranges, but the land itself actually receives much less precipitation – in the range of 1,600 mm to 1,900 mm per annum. The whole of the catchment area of Yodo River is 8,240 km^2, including Lake Biwa (3,848 km^2), of which about half is forest area. Of the other half, about 30 per cent is agricultural land, most of which is paddy fields, and less than 20 per cent is used for housing and other development activities.

The mountain ranges around Lake Biwa and the tributaries of

the Yodo River are generally covered with dense forest, which functions as a natural reservoir as well as a gigantic filter producing high-quality raw water to replenish Lake Biwa and the Yodo River. There are, however, 12 dams (including some under construction, but excluding agricultural irrigation dams) in the watershed for the integrated management of water resources to make up for fluctuations in precipitation (see fig. 3.2). The natural advantages, however, face the challenge of human consumptive activities for scores of kilometres downstream.

The urban, industrial, and agricultural water resource needs of the Kansai Metropolitan Region are supported by the Biwa–Yodo system, including those of Kobe, which is not within the watershed. Water is extracted by various water supply systems from the Biwa–Yodo system. The Yodo River main watercourse serves most of the downstream needs, as shown in figure 3.3. Wastewater is collected by the sewerage networks, treated, and discharged back into the Biwa–Yodo watercourse or directly into Osaka Bay and Seto Inland Sea via smaller watercourses. Street runoff is dealt with in the same way. Thus, discussion of water quality management issues in the region invariably involves the Biwa–Yodo system and the Osaka Bay–Seto Inland Sea receiving water bodies.

The water systems of Lake Biwa, Yodo River, and Osaka Bay are unique. Lake Biwa and Yodo River support extensive urban, industrial, and agricultural activities and Osaka Bay supports maritime activities, all within a compact geographical area of less than 10,000 km^2. The direct distance from the northernmost tip of the Lake Biwa watershed to the southern opening of the Osaka Bay estuary is less than 200 km. Osaka Bay, which is not much larger than Lake Biwa itself, is an ideal harbour supporting the port of Osaka, where significant cargo and passenger maritime activities and some active fishery takes place. Also conspicuous at the south-eastern end of the Bay is the newly constructed island for Kansai Airport.

Lake Biwa Comprehensive Development Project

Lake Biwa, Japan's largest freshwater lake in terms of both its surface area (674 km^2) and its volume (27.5×10^9 m^3) and accommodating 121 rivers of various sizes (over 400 if small streams are included), serves as a source of domestic and industrial water both for the lake catchment area and for the downstream population and industrial centres in the Kansai Metropolitan Region (apart from Nara

Kansai Metropolitan Region

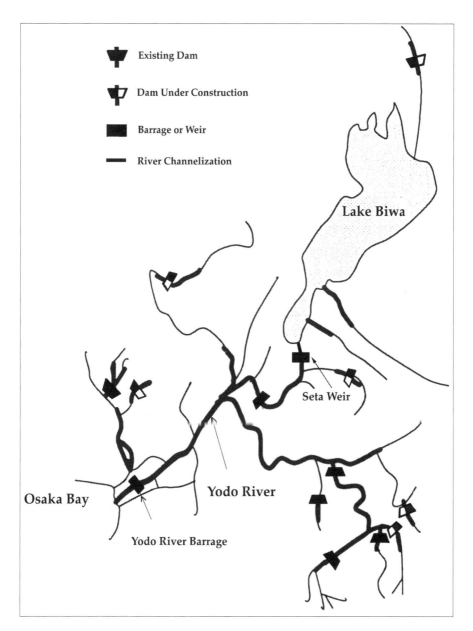

Fig. 3.2 **Water resource development projects in the Lake Biwa–Yodo River region (Source: Biwako–Yodogawa Suishitu Hozen Kikou, 1996, p. 37)**

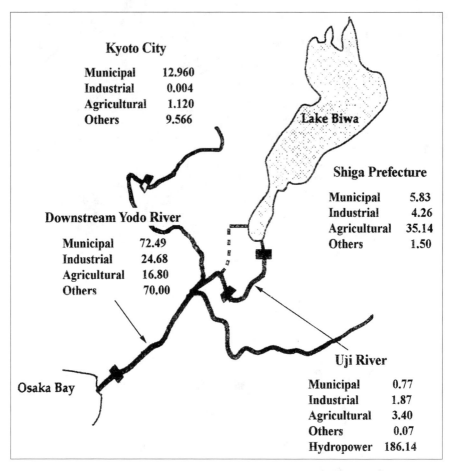

Fig. 3.3 **Allocation of water rights (m³/sec) in the Lake Biwa–Yodo River region (Source: data obtained from Shiga Prefecture Bureau of Lake Biwa Environment, Water Policy Division, 1997)**

and Wakayama). This occurs through the Seta River, the only natural river flowing out of the lake.[2] Within Kyoto Prefecture it is called the Uji River. The Uji River is met by the Kizu and Katsura rivers at the Kyoto and Osaka prefectural boundary, and the downstream stretch from this point is popularly called the Yodo River (see fig. 3.1). The flow contributions to the Yodo River of the Uji, Kizu, and Katsura rivers are, respectively, 64.2 per cent, 18.0 per cent, and 15.0 per cent.

The official designation of the whole of the Yodo–Uji and Lake Biwa water bodies is the Yodo River system. Its annual average flow, its high flow, and its low flow are, respectively, 226.8 m³/sec,

177.6 m^3/sec, and 117.0 m^3/sec. The ratio of high to low flows, 0.52, is the highest among the major river systems in Japan, making Yodo River a very stable source of water. The Yodo River system has many features that make it an excellent water source. These include the beneficial relationship it has with Lake Biwa, which serves as an ideal natural reservoir and regulator of river flow. Other unique characteristics are that the river system gets replenished three times a year by melting snow, by spring rain, and by typhoons. The Yodo River system comprises three river systems with different meteorological features that compensate each other during their individual flow fluctuations. The entire river basin consists of several self-enclosed lowland areas. These features allow for the recapture of highland water for use within the lower areas, making the lowland area in the upper basin of Yodo River a natural underground reservoir that ultimately contributes greatly to the replenishment of the lower stretch of the Yodo River (Fujino, 1970).

Among the distinct phases of Lake Biwa management, the earliest and longest was that of flood and drought management. The paddy farmers in the catchment basin had periodically suffered from flooding and droughts before the 1905 construction of a weir on the Seta River, the only outlet river, which has since moderated the magnitude of their impacts. This phase was followed several decades later by the management of water resources. The Lake Biwa water system, serving downstream needs by way of the Yodo River watercourse, had long been more than adequate. But when Japan began experiencing phenomenal economic growth in the 1960s and downstream demand was difficult to quench with the flow of the Yodo River at the time, the water of Lake Biwa immediately became a hot subject of debate. It was argued that, in view of the projected shortage of water for industrial and municipal uses, the resource value of Lake Biwa water should be exploited further. After lengthy and heated political exchange among the parties concerned, the Shiga and the downstream local and prefectural governments, with the help of the central government, agreed in 1972 to engage in a large-scale water resource development project called the Lake Biwa Comprehensive Development Project (LBCDP).

The basic idea of the project was to allow the discharge of 40 m^3/sec of additional water at times of drought. The corresponding draw-down of the water level was set at 1.5 m below normal. The project comprised water resource development projects, flood control and related water management projects, and compensatory public works projects for the development of catchment land. It was originally a 10-year project. Upon failure to implement the component projects fully by 1982, the project was extended by 10 years, and then for another 5 years, to become a 25-year national project. The project, whose total budget eventually turned out to be 1.9 trillion yen (or some US$19 billion), was finally completed in March 1997.

As a result of this development project, lake water can be released to meet downstream needs as appropriate, and the improvement of the in-flowing river system and the coastal fringe will help alleviate flood damage better. In addition, Shiga Prefecture is much better off in terms both of the economic development achieved during the course of the LBCDP and of future potential, with extensive infrastructure development undertaken thanks to the financial arrangements of the project. The aquatic environment of Lake Biwa, however, has undergone significant change over this period, as will be discussed below.

Water metabolism of the region

Metabolic features of the region

The geographical features of a region are the principal factors determining the configuration of its water system. Metropolitan developments in the Lake Biwa–Yodo River watershed are shaped in the lowland areas, which are separated by plateaus and mountains that obstruct direct and easy links. The road connections between these lowlands are of radial rather than concentric forms, which has allowed the development of metropolises with unique features. The systems of water supply and of wastewater discharge have also been developed more or less independently precisely for that reason.

There are also demographic features characterizing the Kansai Metropolitan Region. One of the most important is that the population density of the region is 2.3 times the national average, and is the highest among the three major metropolitan regions in Japan. The intensity of water use per unit of land as well as of pollution generation is naturally affected by such statistics, though other factors also come into play.

The Osaka–Kobe belt zone, otherwise known as the Hanshin Industrial Zone, was the largest industrial complex until the middle of the Second World War. Industrial output from this region, however, began to decline within a few decades after the war because of the decline in heavy industries such as steel mills and ship-building owing to increasing competition from the newly industrializing nations. Unfortunately, there was no major siting in this region of the thriving post-war industries in Japan such as automobile manufacturing. However, the economy of the Kansai Region is still supported by such industries as iron and steel, chemicals, and textiles (33.7 per cent, compared with 27.5 per cent for Kanto Region and 25.7 per cent for Chubu Region). There is also a relatively greater proportion of manufacturing industries producing household consumption goods such as food, clothing, rubber, plastic and leather items (33.0 per cent, compared with 26.5 per cent for Kanto and 25.7 per cent for Chubu). The proportion of industries producing electrical and electronic equipment or high-tech products such as computers and audio-visual supplies is correspondingly lower (33.3 per cent, compared with 46.5 per cent in Kanto and 48.9 per cent in Chubu). The first two categories of industries are much more water intensive than the third, meaning that the degree of pollution load emitted per unit of production in the Kansai Region may be said to be greater than that in Kanto or Chubu. This general trend, however, does not hold for the Lake Biwa lowland area, where the siting of polluting industries is strictly regulated.

Water supplies

As might be expected, there are many water supply systems in the Kansai Metropolitan Region (87 fully equipped water treatment systems, 292 partially equipped small treatment systems in rural areas, and 63 special-purpose treatment systems including industrial water supplies). Lake Biwa water is supplied to Osaka Prefecture (58%), Kyoto (20%), Shiga (9%), Hyogo (8%), Mie (1%), and Nara (4%) (Biwako–Yodogawa Mizukankyou Kaigi Jimukyoku, 1996, p. 28). Municipal water use has increased and continues to increase as a result of changing life styles demanding greater amounts of water for non-consumption uses such as bathing, showering, and car-washing, and also because of the proliferation of household appliances requiring large amounts of water, such as automatic washing machines and dishwashers (see fig. 3.4). Demand for industrial water use, on the other hand, has been steadily decreasing owing to the change in the nature of industrial activities and to the greater awareness of water conservation.

In principle, responsibility for the management of municipal water

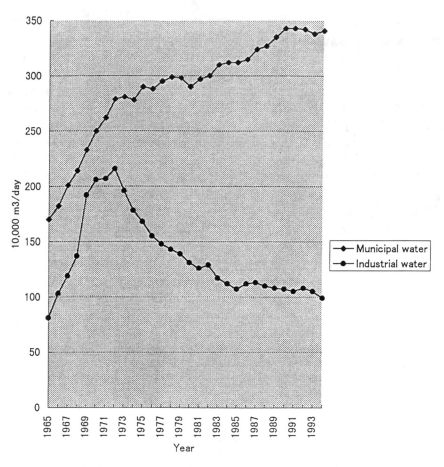

Fig. 3.4 **Trends in municipal and industrial water use in Osaka City and Osaka Prefecture, 1965–1994 (Source: data obtained from Osaka Prefecture and Osaka City)**

supply systems lies with municipal governments in Japan. They need to source their water within their own jurisdictional areas and construct and manage the systems.

The Kyoto water supply system, consisting of four purification plants (including the Keage Plant, which was the first plant in Japan to be installed with rapid sand filters), is almost totally dependent on Lake Biwa water. The Lake Biwa canals (called Biwako Sosui, one completed in 1890 and the other in 1970) provide a total of 23.35 m^3/sec of lake water, much more than the city needs. The quality of the water is that of the southern basin of Lake Biwa where the canal intakes are located. The major concern of the Kyoto purification plants, aside from ordinary concerns about chemical and bacteriological contamination, is the problem in early summer of musty taste and odour, which are caused by certain phytoplanktons in the Lake Biwa water. Wastewater drains into the two tributary rivers to the Yodo River.

Osaka City possesses three large water treatment plants (Niwakubo, Toyono, and Kunijima), which supply four distribution districts. All the plants depend on the Yodo River as their water source, with total water rights amounting to 23.49 m^3/sec or about 203 million m^3/day (Osaka Municipal Water Works Bureau, n.d.). The quality of the water drawn into these water purification plants is significantly affected by Lake Biwa's water quality as well as by the quality of water draining into the Yodo River from its tributaries that run through urban and semi-urban developments. In particular, the problem of musty taste and odour, which first affected the Osaka water supply in 1970, has become of symbolic importance in terms both of consumers' awareness of the deterioration in the quality of Lake Biwa and Yodo River water and of the emphasis by the Osaka City and Prefecture waterworks on the need to upgrade their treatment systems (see, for example, Osaka-fu Suido-bu Suishitu Shikenjyo, 1995). The presence of the Kyoto wastewater treatment plants upstream along the Kamo River also makes it necessary for Osaka City and Prefecture to resort to treatment technologies that produce water of the highest quality at a very high unit cost of production. By the mid-1990s they were finally able to convince the public of the need to increase the water rates and to mobilize financial resources to begin to cope with the problem.

Except for large cities such as Kyoto and Osaka, rapidly increasing demand and limited water sources have made it impossible for individual municipalities to develop and manage their own water supply systems, prompting the development of regional water supply systems managed by the prefectural governments. In the suburban Osaka area, for example, the Prefectural Water Supply Utilities supply treated water at wholesale prices to 39 small and medium-sized municipalities. The prefectural government of Osaka operates three major municipal water treatment plants (Murano, Mishima, and Niwakubo), and two additional water supply systems are currently under construction. A semi-public regional water supply utility called the Hanshin Water Supply Corporation provides water to municipalities in the western part of Osaka Prefecture and the eastern part of Hyogo Prefecture (including the city of Kobe).

The development of industrial water systems in the Kansai Metropolitan Region has its own history. Prior to the installation of industrial water supply systems, industries in Osaka took water straight from rivers or from underground. The rapid growth of industries after the war, particularly in the 1950s and 1960s, led to the extraction of large amounts of groundwater, which began to cause serious land subsidence in the western part of Osaka along the coastline.[3] To prevent further subsidence, groundwater extraction had to be replaced by the provision of water specifically for industrial use. Several water supply systems in Osaka City as well as in Osaka Prefecture are dedicated to supplying industrial water (see fig. 3.5). In Osaka City, four water purification plants supply four service districts with some 271,400 m^3/day. In Osaka Prefecture, two water treatment plants provide nearly 344,000 m^3/day (as of 1995).

Pollution control and wastewater management

The task of reviewing the history of air, water, and soil pollution in the Kansai Metropolitan Region and the measures taken by local governments to overcome it is enormous and beyond the scope of this paper. However, it is important briefly to review the background to the successful pollution countermeasures. One document (World Bank and EX Corporation, 1995, Annex 1, p. 113) makes some important observations about the Osaka City experience:

1. Ground subsidence has been a critical problem in Osaka since the pre-war period. The public and private sectors co-operatively implemented subsidence prevention ordinances ahead of national measures and established a system to implement countermeasures.
2. The City of Osaka used the approach of public and private co-operation. The Osaka mayor prioritized removal of industrial pollution as the most important policy in the city, and designed a strategy for pollution control based on available technology in co-operation with scientists and researchers.
3. The City of Osaka systematised financial support measures, such as the "Osaka City Loan for the Installation of Anti-Pollution Equipment" and the "Purchase System of the Site of Relocated Pollution Source Facilities" for small and medium-sized companies and gave them priority when dispensing financial assistance.
4. With respect to urban wastewater, the city set up guidelines for a basic policy and implemented infrastructure beginning from the pre-war period.

Kansai Metropolitan Region

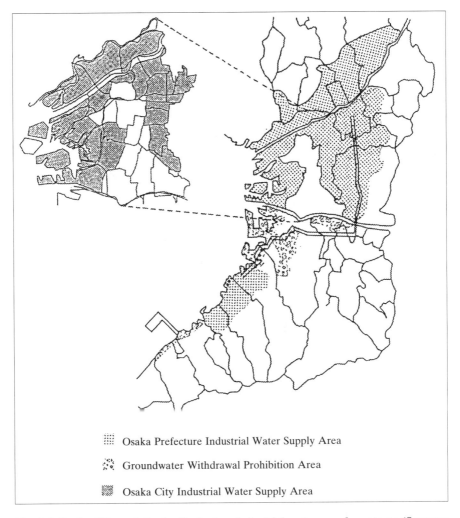

Fig. 3.5 **Osaka City and Osaka Prefecture industrial water supply systems (Sources: Osaka Municipal Water Works Bureau, n.d., p. 35; Osaka-fu, 1996, p. 111)**

It is interesting to note in this respect that the biochemical oxygen demand (BOD) discharged into rivers in Osaka City in 1970 was 573 tons/day, of which about 60 per cent was industrial wastewater, whereas in 1990 the BOD was 160 tons/day, of which less than 20 per cent was industrial wastewater. The general process through which the once serious pollution of air, water, and soil was approached did not differ much among municipalities in the Kansai Region or, for that matter, in other parts of Japan, although the measures taken to deal with specific local issues differed significantly.

Let us look at the development of municipal wastewater systems in the Kansai Metropolitan Region. Osaka, Kyoto, and Kobe have nearly complete service coverage within their jurisdictions, and each metropolis has its own unique features that reflect specific local requirements.

The most notable among the three is the wastewater system of Osaka City. The sewerage service boasts nearly 100 per cent coverage within the city jurisdiction and the cheapest per capita sewer charge in Japan. The current wastewater system developed over a period of more than 100 years, not taking into account the remains of the open sewer constructed in the sixteenth century together with the famous Osaka Castle, part of which is still in use (with suitable improvements). One of the primary reasons Osaka City has placed great emphasis on developing the sewerage system is the fact that the city had been flooded frequently owing to its coastal lowland terrain. The design of the sewerage system therefore had to place great emphasis on flood control, resulting in the use of what is called "the combined sewer system."

The combined sewer system allows not only raw sewage but also rainwater to be collected and delivered to treatment plants, thus requiring larger sewer pipes and plants. However, when there is heavy rain, even the large wastewater treatment plants cannot cope. The excess has to be drained to watercourses as "overflows," causing pollution of watercourses by the effluent from the sewerage treatment plants themselves. Improvement of the existing combined sewer system is the major problem facing the city (only 1.1 per cent of the whole sewerage system is a separate system that allows stormwater simply to be discharged without being mixed with raw sewage) (Osaka Municipal Sewage Works Bureau, 1993).

As for the city of Kyoto, a feasibility study for the construction of a sewerage system was first initiated in 1894 but no serious construction activity was undertaken until 1930 when it was integrated into public works measures to counter unemployment. The project was not pursued with much vigour owing to Japan's financial state prior to the war. It was only in 1934 that the first wastewater treatment plant came into operation, primarily to deal with heavy pollution of the Kamo River. The Toba wastewater treatment plant, the third-largest plant in Japan today, was brought into operation in 1936, and the city gradually expanded its coverage. Overall service coverage has now reached 97 per cent. The sewerage system in Kyoto is characterized by a combination of combined and separate sewers, meaning that some 40 per cent of wastewater may have to be released into watercourses after heavy storms. Four municipal wastewater treatment plants cater to the city's needs, each giving specific consideration to treating wastewater discharges from various industries, including typical Kyoto industries such as textile dyeing and *sake* distillation.

The wastewater system in Kobe, whose systematic development began in 1951, is very different from both the Osaka City and the Kyoto City systems in that it is an almost totally separate system. Nearly 1.2 million m^3 of wastewater are treated by seven plants, four of which are located on the shores of the Seto Inland Sea. The terrain of the city consists of reclaimed coastal land, hillsides, and an upper plateau region, which makes for separate and independent development of individual systems. It is to be especially noted that serious damage was caused to the sewer network as well as to the treatment plants by the Hanshin-Awaji earthquake of 1995. The current major concerns are rehabilitation of the damaged facilities, development of technologies for the reconstruction and improvement of facilities to withstand earthquakes of major magnitude, as well as justification for upgrading the current system to accommodate nutrient removal in order to protect the Seto Inland Sea from further eutrophication.

The wastewater systems that serve areas other than the three city jurisdictions are also of major concern as regards the protection of the Yodo River and Osaka Bay. Many of the smaller municipalities were unable to raise adequate financial resources to develop their own wastewater systems and have been brought together into regional wastewater systems. There are some 20 regional wastewater systems in Kansai Region at various stages of development. In the case of Osaka Prefecture, 12 regional wastewater systems serve some 42 municipalities at various levels of coverage, ranging from nearly 100 per cent for 4 municipalities to less than 50 per cent for 17 municipalities (as of 1995). The average service coverage for the whole of the prefecture is 63.7 per cent. In the case of Kyoto Prefecture, two regional treatment plants are of direct significance to the protection of Yodo River water quality. The Rakusei wastewater treatment plant was brought into operation in 1979 and the Rakunan plant in 1986. The overall extent of service coverage was reported to be 73.6 per cent in 1995.

The wastewater system to protect Lake Biwa is another important subject (fig. 3.6). In Shiga Prefecture, household wastewater has been

Kansai Metropolitan Region

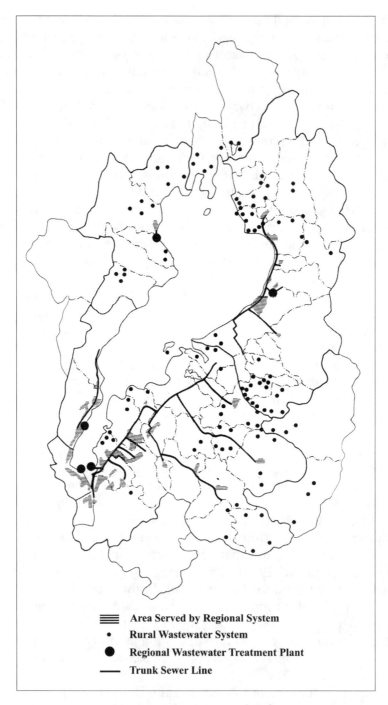

Fig. 3.6 **Wastewater systems in Shiga Prefecture (Source: Shiga Prefecture)**

and still is a major polluter of the lake. It is treated by a variety of sewerage systems. Although municipal sewerage systems, including the regional systems, will eventually cover more or less the entire flat part of the watershed, the current coverage is only 47 per cent of the population (50 per cent of the population eventually to be served). They are equipped with advanced treatment facilities to remove nutrients in order to protect Lake Biwa from eutrophication.

There are also small-scale sewerage systems for agricultural communities and a variety of on-site sewerage systems collectively called "the Jokaso system". Some are equipped with advanced treatment capability. Together, they currently serve some 17 per cent of the population. That leaves some 36 per cent of the population still not served by conventional flush toilet systems. Excreta collected from these households are transported to 12 night-soil treatment systems, which are all equipped to treat not only the organic content but also nutrients.

The regional sewerage systems are to be gradually expanded to integrate in their coverage many of the agricultural community systems and on-site facilities. By around 2010 these regional systems together with a municipal system for the city of Otsu will serve nearly 90 per cent of the population. The remaining 10 per cent will still have to be served by the agricultural community systems.

Water quality issues in the Kansai Metropolitan Region

Upstream–downstream relationships

The water quality profile of the Lake Biwa–Yodo River–Osaka Bay system is quite complicated because there are many different kinds of upstream–downstream relationships. The most obvious relationship is between Lake Biwa, with extensive watershed development, upstream and the great metropolitan region downstream. In the upstream Lake Biwa watershed, infrastructure development through the LBCDP has literally transformed water-use practices over recent decades, in that the new infrastructure for water intake and transmission has made the lake water much more attractive than river water or groundwater because of its abundance, stability, and good quality. Nearly 80 per cent of municipal and agricultural water supplies now come from the lake and nearly all of it is returned to the lake, either with treatment (mostly in the case of municipal supply) or without treatment (mostly in the case of agricultural return flow).

There are other more localized upstream–downstream relationships throughout the Lake Biwa–Yodo River system. For example, there are many problematic relationships between upstream wastewater effluent discharge points and downstream water intake points, particularly along the middle and lower stretches of the Yodo River. Within the short stretch of 30 km or so between the Hirakata confluence points of two major tributary rivers of Yodo River and the river mouth at the Osaka Bay inlet, there is a concentration of discharge and intake points (see fig. 3.7). Upstream of some major water intake points, there are treatment plants discharging wastewater, sometimes in large quantities, as well as streams and channels discharging contaminated water from areas undergoing extensive urban development. Obviously the water withdrawn for treatment must undergo an extensive purification process before being distributed for consumption.

This upstream–downstream relationship may be examined in more detail in terms of the institutional and political context in two cases, the first involving the Shiga Prefecture and downstream governments as a whole, and the second involving Kyoto and Osaka.

In the first case, the conflict is between downstream expectations and the upstream mandate. To put it simply, the downstream local governments require (as they may be entitled to because of their financial contribution to the LBCDP) that the upstream Shiga Prefecture should keep the lake water as clean as possible. The quality of water the downstream users get is greatly dependent on the quality of the Lake Biwa water. For the Shiga Prefecture, on the other hand, keeping the lake water clean is more a mandate than an obligation, since the lake was originally clean and for centuries the prefecture has been fostering this great natural asset. The mandate, however, does not mean that the prefecture can pay to maintain or restore the quality of the lake water, which leads naturally to the thinking that the financial burden has to be shouldered by others as well, including and especially the downstream users. However, it is difficult to decide how clean is clean enough and who is to pay how much to get the lake water to a desirable quality. The LBCDP expenditure on environmental conservation has not fulfilled the mandate, in the sense that the costs expended through the LBCDP arrangement, which includes national and downstream government contributions, have been far from realizing a clean lake.

The second case involves an interesting but difficult issue regarding the Yodo River water quality between Kyoto, upstream, and Osaka, downstream. For the Osaka water suppliers, it is crucial that the

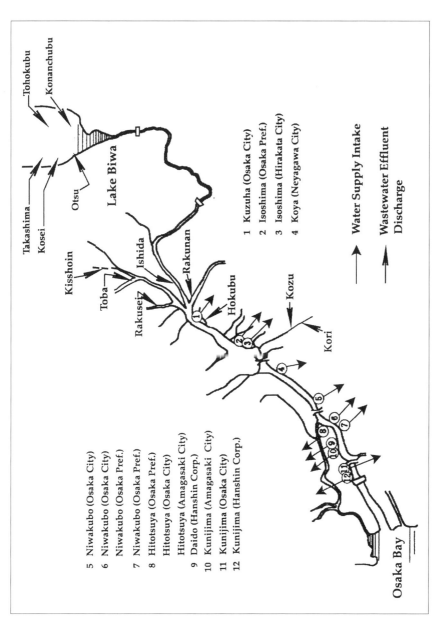

Fig. 3.7 Water supply intakes and wastewater effluent discharge points along the Lake Biwa–Yodo River watercourse (Note: not to scale. Source: based on figure in Osaka-fu Suido-bu Suishitu Shikenjyo, 1995, p. 4)

quality of the Yodo River be kept reasonably high (at least to the level of acceptable raw water quality, both technically and psychologically). Otherwise, not only does the cost of treating water become prohibitively expensive, but there is a strong psychological impact on Osaka citizens if they think that their raw water is heavily contaminated by Kyoto wastewater. Osaka already expends great effort to treat its raw water because the lake water and the water in the tributary rivers to the Yodo River upstream of the Kyoto wastewater discharges are already significantly problematic. The lake water is eutrophic and the river water is polluted by urban and agricultural discharges further upstream. For Kyoto, on the other hand, it is important to keep the quality of effluent discharges free of residual pollution over and above the level the downstream water users find acceptable, while keeping the cost of treatment reasonable. The downstream water supply systems have been asking Kyoto City to reduce its pollution impact further.[4]

Upgrading of wastewater management systems

The above discussion brings up the question both of the manageability of pollution of receiving water bodies and of the treatability of contaminated raw water for drinking. This subject involves several complicated issues.

First, control of point source pollution in the upstream regions of the Lake Biwa–Yodo River system is still inadequate, with some lagging far behind the metropolitan districts of Kyoto, Osaka, and Kobe cities. In the Lake Biwa watershed, for example, only 54 per cent of the whole of the prefectural population is served by public sewerage systems. The overall extent of coverage in municipalities in the Lake Biwa–Yodo River watershed other than the three metropolises is around 60–70 per cent. The rest of the population is served by sanitary systems with less complete removal of organic matter and nutrients. In other words, the nature of the upstream–downstream relationship is bound to be affected by progress in the provision in future of sewerage services upstream.

Second, the control of non-point source pollution is becoming a subject of greater concern year by year. Two types of non-point pollution need close examination: the first is of agricultural origin in the Lake Biwa watershed; the second is of urban origin, particularly along the Yodo River portion of the watershed.

Agricultural non-point pollution is of concern because of its large contribution of organic substances as measured in terms of chemical oxygen demand (COD, 49 per cent), total nitrogen (TN, 56 per cent), and pesticides. As the point sources of pollution become more and more stringently controlled, the agricultural non-point share becomes greater, though its control is still not technically or financially feasible. The combined impact of eutrophication and toxicity on biota in the estuary waters of the lake may further disturb the ecosystem, as reflected in phytoplankton blooms and loss of biodiversity.

Non-point pollution from urban sources contains toxic organic and inorganic substances mostly in the form of runoff from streets. They are of increasing concern because of the combined risks to health from long-term uptake through drinking water supplies. For example, the concentration profile of organophosphoric acid triesters in the Osaka metropolitan watercourses and along the Yodo River (Fukushima et al., 1992; see fig. 3.8) attests to the fact that the level of contamination of urban streams along the Yamato River, for example, is quite serious.[5] This situation will be difficult to contain unless the sewerage programme is closely linked with control of the use of chemicals, street cleansing, traffic and transportation planning, as well as land-use control, all to prevent street runoff being discharged directly into the receiving water bodies.

Fukushima and Yamaguchi (1992) also observed that the concentration distribution of pesticides such as molinate, fenitrothion, and isoprothiolane correctly reflects the kinds of activities the chemicals are used for upstream of the survey points in the watershed system. Thus, molinate is found mostly in the main watercourse of the Yodo River itself because of its widespread use in the paddy fields in the Lake Biwa watershed; fenitrothion is found in high concentrations in suburban watercourses and not in the main Yodo River watercourse because of its extensive use in horticulture and household pest control; and isoprothiolane is found both in the Yodo River system as well as in urban streams because of its popularity as an insecticide both for paddy pest control and for the protection of urban greenery such as golf links, parks, and other recreational areas (fig. 3.9).

As already mentioned above, the Lake Biwa–Yodo River system receives significant amounts of wastewater effluent from large wastewater treatment plants located along its watercourse. Although these wastewater treatment plants either have been upgraded or are in the process of being upgraded to produce effluent of higher quality (to the extent of producing effluent of a quality close to that of drinking water in some cases, as is the case with the Lake Biwa re-

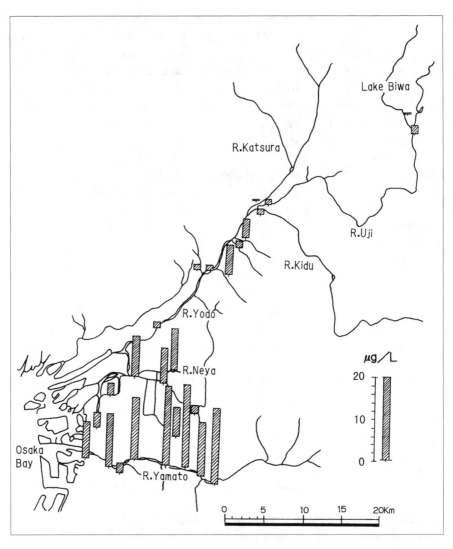

Fig. 3.8 **Concentrations of organophosphoric acid triesters in the Yodo River** (Source: Fukushima et al., 1992, p. 274, fig. 2)

gional wastewater treatment plants), the overall situation is still far from adequate and these effluents are still major sources of pollution in the water system as a whole. There are also ongoing programmes for upgrading facilities to treat combined sewer outflow at times of severe storms by storing the initial outflow portion in temporary underground storage facilities that will range in length up to several kilometres and in diameter up to several metres.

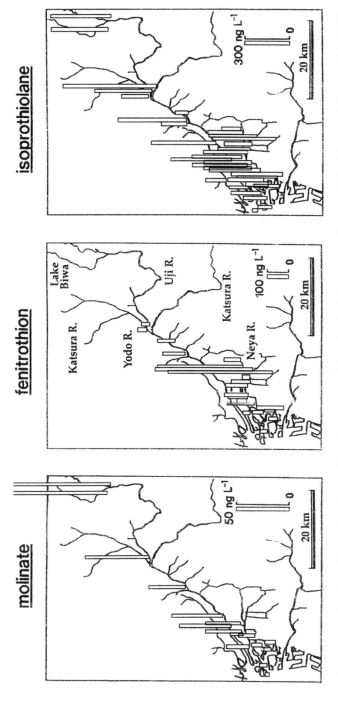

Fig. 3.9 Distribution of agrochemical residues in the Lake Biwa–Yodo River water (Source: Fukushima and Yamaguchi, 1992)

The water quality of Osaka Bay

The water quality of Osaka Bay throughout the post-war era until the late 1970s was extremely poor. Unlike Lake Biwa, the Bay served no purpose in terms of water supplies and the deterioration of water quality had long been regarded as a necessary evil of the industrialization of the Keihanshin Industrial Belt. The situation began to improve as industrial discharges of noxious pollutants began to be severely curtailed and sewerage services in the upland began to have an effect in terms of reducing household waste load inputs to the Bay.

A brief review of the history of water quality management in the Bay reveals that control of eutrophication of marine environments became a major concern in national environmental programmes after the first sighting of a large-scale marine red tide in the early 1970s. The Soryo Kisei, or total waste load control policy, has been adopted for Seto Inland Sea since the Special Measures Act for the Environmental Protection of Seto Inland Sea was promulgated in 1973. For the Kansai Metropolitan Region as a whole, the control of eutrophication both of Osaka Bay and of Seto Inland Sea is of great concern. The total waste load reduction policy requires that industries with a daily waste load of more than 50 m^3 discharging into the designated receiving body of water systematically reduce their nutrient inputs. For example, the concentration of nitrogen (N) and phosphorus (P) in the wastewater discharged by such industries must be less than 120 mg/litre (the average daily concentration must be below 60 mg/litre) and 16 mg/litre (8 mg/litre), respectively, to protect Seto Inland Sea from further eutrophication.

The total waste load control policy requires not only industries but also municipal waste treatment facilities to upgrade treatment plant capability to further reduce the nutrient content in the plant effluent. Because the eutrophic condition of the Bay has not improved much since the institution of the programme, it is envisioned that greater investments will be required for municipalities to upgrade their nutrient removal capability in the next round of policy review. Based on the Special Measures Act of 1973, the relevant prefectural governments must develop a plan and revise it every five years. The first such plan was developed in 1978 and, in general, they have to resort to multiple regulatory programmes and facility construction programmes to pursue the required waste load reduction. For example, the Hyogo prefectural government revised its plan in 1993 (Hyogoken, 1996) in such a way that the plan would be compatible with the implementation schedules stipulated in other programmes such as those for sewerage construction and for total waste load reduction plans.

As for the pollution profile with regard to synthetic chemicals, the distribution and concentration levels found indicate that the situation in Osaka Bay is significantly affected by the upstream Yodo River as well (Fukushima, 1996).

Control of Lake Biwa eutrophication

The first serious sign of a deterioration in Lake Biwa water quality came suddenly, in the form of a large-scale red tide along the eastern coastline of the northern basin in early 1977. The sighting of this phytoplankton, *Uroglena Americana*, was quite a shock as the northern basin had been, up till then, believed to be nearly pristine. The red tide has since been sighted almost every year. Prompted by the red-tide incident, the Shiga Prefecture enacted the Eutrophication Control Ordinance of 1980 to ban the use and sale of phosphorus-containing synthetic detergents. A wide array of control measures has been introduced over the decades for improving Lake Biwa water quality. Already by the mid-1970s it was becoming difficult for large-scale industries to get away with discharging polluting wastewater into watercourses leading to the lake, thanks to the Water Pollution Control Law of 1970, which had stringent punitive provisions. With the enactment of the Eutrophication Control Act of 1980, the regulatory provisions for industries became even more precise and stringent, with additional controls on nutrient discharges. Wastewater from smaller-scale industries has been progressively brought under control, although very small industries (whose discharges are insignificant in comparison with total industrial discharges) are yet to be regulated fully. Despite all the environmental programmes, typical water quality indices, such as COD, total phosphorus (TP), and total nitrogen (TN), reveal that the improvement in lake water quality has not been very impressive; indeed, figure 3.10 shows a worsening trend in COD in recent years.

What about trends in water quality over the much longer term? Water transparency at specified sites and bottom dissolved oxygen in the northern lake are two indicators of longer-term trends that happen to be available. They show that Lake Biwa water quality had slowly been deteriorating even before the red-tide incident of 1977. In addition, there are other more subtle indicators of what is taking

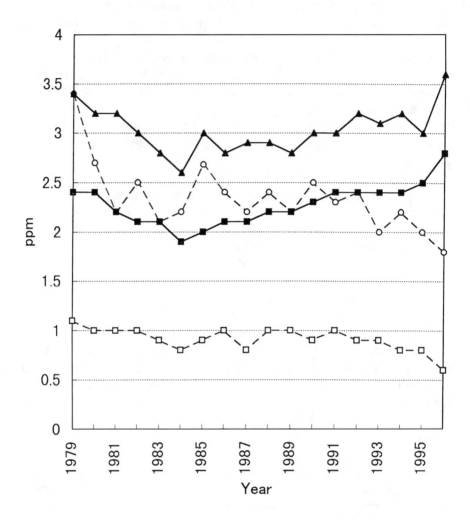

Fig. 3.10 **Trends in Lake Biwa water quality: TP and COD, 1979–1996 (Source: Shiga-ken, 1997)**

place in the lake ecosystem. For example, figure 3.11 shows that there is a worrying change in the dominant species of nuisance-causing phytoplankton in the lake. Putting all the indicators together, it can be noted that the deteriorating trend was accelerating up until the mid-1980s and, although the rate has since slowed down, the trend itself has not been reversed yet.

Are there good prospects of reducing polluting inputs into the lake? The regulation of polluting industries and the provision of sewerage to households and commercial establishments are the two main point source management strategies. Time (how soon?) and the mobilization of financial resources (how much?) are the basic considerations affecting progress now that the special budgetary provisions for environmental projects in the LBCDP have been terminated.[6]

Concern about the control of waste load discharge has been gradually shifting in the past decade or so from point sources of pollution to non-point sources of pollution. The four major sources of non-point pollution are: (1) rainwater, (2) forest and field runoffs, (3) paddy runoff, and (4) urban runoff. Not only will the development of a comprehensive non-point source control system be expensive, but it will also require legal and institutional measures yet to be elaborated. This is particularly so with wet weather non-point sources of pollution, or stormwater runoff into the lake. Moreover, financial resources are likely to be tied up in point source control for some time.

Elaboration of control programmes for non-point pollution from both urban and agricultural sources has just begun. Paddy field runoff under dry weather conditions may be managed very effectively by a combination of structural and non-structural means. The current trend in policy is to regard individual paddy fields as point sources rather than non-point sources under dry weather conditions. Proper management of irrigation water and a reduction in the wasteful use of fertilizers and pesticides are the keys to successful control of dry weather runoffs. The promotion of agricultural best management practice and technological developments for the more efficient use of resources will also be necessary.

Summary and emerging issues

Summary

As described above, the development of the water supply, wastewater management, and the aquatic ecosystem environment in the

Kansai Metropolitan Region

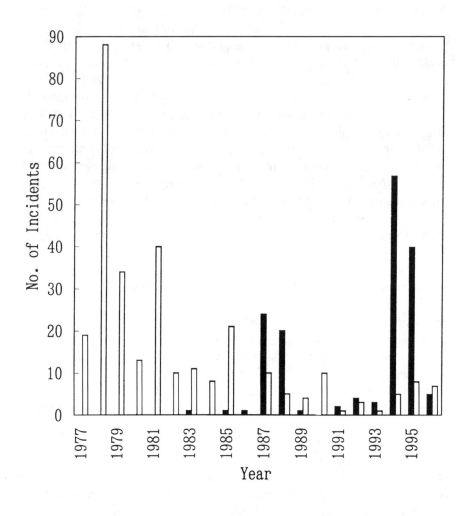

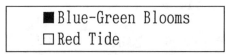

Fig. 3.11 **Trends in algal bloom incidents in Lake Biwa, 1977–1996 (Source: Shiga-ken, 1997)**

Kansai Metropolitan Region has been very much affected by, and had a great deal of impact on, the shaping of the Lake Biwa–Yodo River–Osaka Bay water system. In the process, many different physical systems and managerial approaches have evolved in the region, both in the metropolitan districts of Kyoto, Osaka, and Kobe and in the suburban municipalities outside the metropolitan districts and the urban, semi-urban, and rural areas within the watershed of Lake Biwa. The overall situation is that the problems they face are all quite different, and the need to invest in the development of wastewater-related infrastructure will continue for these areas. Specifically, further increases in sewerage coverage are vital for Shiga Prefecture not only because of the need for greater curtailment of Lake Biwa eutrophication but also because of the need to upgrade basic environmental infrastructure provisions for the rapidly urbanizing Lake Biwa watershed.

Emerging issues

Additional issues of importance may be broadly categorized into: sustainable water use, integrated watershed management, and the attainment of a sound ecosystem. Some elaboration may be useful on these points.

Sustainable water use

The underlying major objective of Lake Biwa–Yodo River management in recent decades, with particular reference to meeting the growing water needs of the Kansai Metropolitan Region, has always been to achieve sustainable water use. In a narrow sense, this seems to have been the mandate of the individual municipal and prefectural governments engaged in water and wastewater management schemes in this region. The major hurdle to achieving sustainable water use, or having access to enough water, has been overcome in principle through the completion of the LBCDP, except that it pertains only to sustainable water quantity, not quality.

As far as the downstream municipalities are concerned, the sustainable use of Lake Biwa water that meets the quality requirements for their water supplies has a lot to do with the cost of water purification. For example, the City of Osaka has been and will be prepared to pay for the advanced treatment of its drinking water from the Yodo River, but it would not be prepared to pay to improve (marginally at best) the quality of Lake Biwa water as a whole, because it would simply not be cost effective. The discussion on the sustainable use of water in respect to both quantity and quality therefore has to be pursued on the basis of a much broader definition of sustainability, which brings us to integrated watershed management.

Integrated watershed management
The management of water quantity in the Lake Biwa–Yodo River system has been integrated in the sense that water rights for extraction from the system have been strictly regulated, under the jurisdiction of the Ministry of Construction. The LBCDP was conceived to generate additional water at times of predicted water shortage so as to be able to augment the limited water rights. With the completion of the LBCDP, together with the integrated system of reservoir regulation within the watershed, the integrated management of water quantity has been further improved. Unfortunately, the integrated management of water quality has not been fully addressed in the Lake Biwa–Yodo River–Osaka Bay water system.

The management of water quality within a system as extensive and sophisticated as the Lake Biwa–Yodo River–Osaka Bay water system may be pursued in many different ways and at very different levels of integration. The least onerous system for the individual municipalities would be to pursue their own water quality programmes independently while keeping track of trends in the overall quality of the Yodo River as a whole, with reference to the established ambient water quality standards. This is the level of integration that the Lake Biwa–Yodo River–Osaka Bay water system has currently reached, and it is working fairly well in that water quality has been improving steadily in terms of conventional point-source quality parameters, i.e. meeting effluent discharge and ambient water quality standards. The problem is that this level of integration will not be adequate to address issues of growing concern such as human and ecosystem risks involving, for example, non-point sources of pollution and the degradation of ecosystem integrity. An integrated and basin-wide strategy needs to be developed and followed by all the members of the water system in a much more coordinated way. Provision for such a system has so far been quite limited.

Currently, three associations deal with issues of common interest to those concerned with Lake Biwa and Yodo River management. The Yodo River Water Pollution Control Consultative Association (Yodogawa Suishitu Odakuboushi Renraku Kyougikai) was formed in 1958 with the aim of enhancing water quality conservation, sewerage planning, and water quality monitoring. The association, which now has 25 members, has also established an emergency communication system in the event of accidents affecting water quality of the supply sources.

The Yodo River Water Quality Consultative Association (Yodogawa Suishitu Kyogikai) was formed in 1965 by the major water supply systems withdrawing raw water from the Yodo River. The association undertakes activities such as monitoring and research on the quality of supply sources, petitioning the national government, the upstream governments (prefectural and municipal governments in Kyoto and Shiga), and other agencies on matters related to upstream–downstream issues, consultation on the siting of the discharge outlets of pollution control facilities upstream of water intake points, strengthening of the emergency system in the event of raw water contamination, and promoting water pollution control.

The Lake Biwa–Yodo River Water Quality Conservation Organization (Biwako Yodogawa Suishituhozen Kikou) was established in 1993 to enable local governments to collaborate on engineering restoration measures to improve the water quality of Lake Biwa and the Yodo River. Activities undertaken by the organization include research and development on suitable water quality management technology, the promotion of water quality improvement projects, the collection, compilation, and processing of water quality data and information, and the promotion of public involvement in the beautification and amenity enhancement of rivers and streams (Mizutani, 1993).

Unlike water authorities, these associations can achieve little in terms of the development of a policy that encompasses the multiple government jurisdictions. There is emerging interest in creating a body with the mandate and resources to pursue integrated policies and programmes for the better management of the Lake Biwa–Yodo River environment, but there appears to be a lack of political commitment to pursuing the development of such an institution. The reality is that the current system of management of the Lake Biwa–Yodo River water system has resolved the water quantity issue but water quality concerns vary widely among the local governments. More rigorous study is needed to clarify what is meant by the integration of watershed management policy and the efficient management of water quality with respect to this unique lake–river–bay water system.

The attainment of a sound ecosystem
The LBCDP has not changed the basic physical structure and capabilities of the water management systems either of Osaka, Kyoto, and Kobe cities or of the surrounding municipalities. This is because it involved physical works pertaining basically to facilities related to Lake Biwa and not to downstream facilities, not to mention that it was basically a water quantity project, though significant investments related to water quality have been made for Lake Biwa and its watershed. On the other hand, in the course of the 25-year period of the LBCDP, the issues of lake water quality and ecosystem integrity have come to the fore and become the crucial concerns, particularly for Shiga Prefecture.

The management of Lake Biwa water quality is about to enter a new phase. This phase will involve increased point and non-point source control measures aiming to reduce waste loads further, and the introduction of control measures to achieve a greater degree of integrity of the lake ecosystem, accompanied by improved lake water quality in terms of parameters other than COD, TP, and TN. Emerging features of the new phase include: (1) rezoning to increase protected watersheds, (2) the restoration of ecotones (the transitional zone between littoral and aquatic environments), including the once reclaimed attached (or adjoining) lakes, and (3) the integrated management of priority watersheds.

Rezoning to increase protected watersheds will probably involve a lengthy political process. The major issue will be devising suitable economic incentives such as compensatory payments to existing land owners, since almost all watershed land is privately owned and productive. As for ecotone restoration, there will have to be close collaboration between the agricultural and urban sectors with respect to infrastructure development or redevelopment involving coastal regions of the lake and numerous watercourses. The key to integrated management of priority watersheds is to allow greater collaboration between different sectors of government (such as flood control, irrigation, and sewer systems) in the management of both water quantity and quality, which have hitherto been independently pursued.

Concluding remarks

Although nobody disputes the need to make the transition from an era of water resource development to a new era in which the achievement of ecosystem integrity is the primary objective, this transition demands some fundamental changes in thinking as well as in the approach of long-range government planning. In the days of planning for infrastructure development, whether for water resource development or for improvement in basic public needs such as sewerage, the approach was to set clear quantitative targets, usually on the basis of a political decision, and then mobilize the necessary resources to realize those targets. When assessing the achievement of the targets, the issue was how far behind implementation had been for reasons of a lack of resource mobilization or of capability. Now that the objective is to attain ecosystem integrity, the approach to planning will have to be entirely different. Not only are the targets unclear and difficult to describe quantitatively, the process leading to achievement of the goal will be an evolving one. In other words, determining how sound is sound enough for the ecosystem under consideration must take into account the values of the public, and these values will have to be continually reassessed.

Notes

1. Kansai Metropolitan Region includes the Shiga (the Lake Biwa area) and Kyoto Prefectures to the north-east of Osaka, the Hyogo Prefecture to the west, and the Nara and Wakayama Prefectures in the central and western parts of the Kii Peninsula south of Osaka. The Lake Biwa–Yodo River system does not provide water to the Nara and Wakayama Prefectures, though the area of its coverage is almost synonymously regarded as the Kansai Region.
2. Two canals connecting to Kyoto City and an intake channel (located immediately downstream of the lake off Seta River) to Uji hydropower dam site are additional water outlets from Lake Biwa.
3. The lowland area along the coast of Osaka Bay was a vast wetland before the land was drained some centuries ago.
4. An integrated system of water quality management would be the ideal. However, it is generally impractical to coordinate the management of water and/or wastewater systems that have evolved independently against different institutional and political backgrounds, as in the case of the Yodo River. The interaction between downstream water supply systems and the Kyoto treatment plant at Toba is an interesting example of the coordinated operation of water and wastewater systems in the Yodo River. The Toba plant was able to reduce the concentration of ammonium nitrogen during winter months beyond the regulatory standard at the request of the downstream water supply agencies. These were suffering from the need to increase the chlorine dosage at their water treatment plants, which results in the enhanced concentration of tri-halomethanes or carcinogenic agents.
5. Organophosphoric acid triesters such as trialkyl and triaryl esters of phosphoric acid are used as flame retardants, plasticizers, hydraulic fluids, and lubricant

additives. Total production seems to have increased remarkably since the 1980s, and was estimated to be about 15,000 tons in 1985. The main concern as regards environmental pollution originates from the structural similarity to many organophosphorus pesticides. In the absence of any sound data on the relationship between structure and toxicity, it may be safe to speculate that the environmental behaviour and the effects of these compounds are similar (Fukushima et al., 1992).
6. Though the LBCDP did not include lake water quality as a primary ingredient in the original plan, a significant amount of investment was budgeted to undertake environmental component projects when the LBCDP was extended for 10 years in 1982.

References

Biwako–Yodogawa Mizukankyou Kaigi Jimukyoku [Executive Office of the Lake Biwa–Yodo River Water Environment Conference]. 1996. *Biwako–Yodogawa wo Utuskushiku Kaeru Tameno Shian* [Provisional Plan for the Attainment of Beautiful Lake Biwa and Yodo River] (in Japanese).

Biwako–Yodogawa Suishitu Hozen Kikou [The Lake Biwa–Yodo River Water Quality Conservation Organization]. 1996. *Biwako Yodogawa no Suishitu Hozen* [Water Quality Conservation of the Lake Biwa–Yodo River Region] (in Japanese).

Fujino, Y. 1970. "Yodogawa no Mizushigen Kaihatu to Suishitu Hozen [Water Resources Development and Water Quality Conservation of Yodo River]," *Doboku Gakkaishi* [Journal of Japan Society of Civil Engineers], Vol. 55, No. 2, pp. 13–18 (in Japanese).

Fukushima, M. 1996. "Mizukankyo ni okeru Jinko Yuukikagoubutu no Doutei to Seibutsunoshuku Kiko ni kansuru Kenkyu [The Behaviour of Synthetic Chemicals and Bioconcentration Mechanisms in Water Environment]," Ph.D. thesis, Kyoto University, Japan.

Fukushima, M. and Yamaguchi, Y. 1992. "Biwako Yodogawa-Suikei, Osaka-shinai Kasen ni Miru Noyaku-osen no Tokucho [The Pollution Characteristics by Agrochemicals of the Lake Biwa–Yodo River Watershed and Osaka City Rivers]," *Kankyo Gijyutsu* [Environmental Technology], Vol. 21, No. 4, pp. 271–276 (in Japanese).

Fukushima, M., Kawai, S., and Yamaguchi, Y. 1992. "Behavior of Organophosphoric Acid Triesters in Japanese Riverine and Coastal Environment," *Water Science and Technology*, Vol. 25, No. 11, pp. 271–278.

Hyogo-ken [Hyogo Prefectural Government]. 1996. *Hyogo-ken Kankyo Hakusho* [Hyogo Prefecture White Paper on the Environment]. (in Japanese).

Mizutani, M. 1993. "Suidou Suigen no Suishituhozenn [Water Quality Conservation of Supply Sources]," *Toshimondai Kenkyuu* [Urban Issue Studies], No. 45, pp. 106–120 (in Japanese).

Nakamura, M. 1995. "Lake Biwa: Have Sustainable Development Objectives Been Met?" *Lakes and Reservoirs: Research and Management*, Vol. 1, pp. 3–29.

Nakamura, M. and Akiyama, M. 1991. "Evolving Issues on Development and Conservation of Lake Biwa–Yodo River Basin," *Water Science and Technology*, No. 23, pp. 93–103.

Osaka Municipal Water Works Bureau. n.d. *Water Supply System in Osaka*.

Osaka Municipal Sewage Works Bureau. 1993. *For a Cleaner, Healthier Environment: Osaka Sewage Works.*

Osaka-fu [Osaka Prefectural Government]. 1996. *Osaka-fu Kankyo Hakusho* [Osaka Prefecture White Paper on the Environment] (in Japanese).

Osaka-fu Suido-bu Suishitu Shikenjyo [Osaka Prefectural Government, Water Quality Control Center of the Department of Waterworks]. 1995. *Biwako to Yodogawa no Suishitu* [Water Quality of Lake Biwa and Yodo River] (in Japanese with English summary).

Shiga-ken [Shiga Prefectural Government]. 1997. *Shiga-ken Kankyo Hakusho* [Shiga Prefecture White Paper on the Environment] (in Japanese).

World Bank and EX Corporation. 1995. *Japan's Experience in Urban Environmental Management.* Prepared by the Metropolitan Environment Improvement Program of the World Bank.

4

Water management in mega-cities in India: Mumbai, Delhi, Calcutta, and Chennai

Rajendra Sagane

Introduction

Addressing infrastructural deficiencies in Indian mega-cities is a daunting task and large and metropolitan cities present a dismal picture today. The deficiencies will become even more pressing with continuing growth in future. The constraints inherent in the present scenario include insufficient water resources, an inadequate organizational set-up, a lack of funding resources, and inadequate management and technical skills. Another disturbing factor is the lack of political support from government to take timely action to solve the problems.

India's mega-cities are the generators of national wealth. Urban India contributes more than 50 per cent of the country's GDP while containing only 27 per cent of its total population. India's large towns, however, suffer from several problems that accompany economic development, namely high population densities, unplanned settlements, slums, traffic pollution, environmental decay, mounting poverty, unemployment, and social tensions and unrest. Consequently, the urban environment (land, water, and air) is deteriorating in many towns.

The urban scenario in India

India had a population of 658 million in 1981 and 844 million in 1991, and projections for 2001 and 2025 are 1,000 and 1,500 million, respectively. According to the 1991 census, there are 3,768 urban agglomerations or towns with a total population of some 217 million, which is about 25.75 per cent of the country's total population of 844 million. There are 23 metropolitan towns with a population of over 1 million each, compared with 12 in 1981. These 23 cities account for roughly one-third of the country's urban population and one-twelfth of the total population of the country. Though urbanization contributes to the growth process, by and large this positive aspect is often overshadowed by a deterioration in the physical and environmental quality of life in urban areas on account of the wide gap between demand for and supply of essential services and infrastructure.

Indian mega-cities

Population trends

Based on population, six Indian cities fall into the category of mega-cities: Mumbai, Calcutta, Delhi, Chennai, Hyderabad, and Bangalore (fig. 4.1).[1] One-fifth of India's total population lives in these six cities. The population in these cities saw an increase of 33 million during the four decades from 1951 to 1991 (an average of 8.2 million per decade) and it is anticipated that future growth from 1991 to 2001 will be 18.4 million. However, the population of these six mega-cities as a percentage of the total population of India has remained more or less constant: in 1951 it was 20.1 per cent; in 1991 it was 21 per cent; and in 2001 it is projected to be 21.53 per cent.

The slum population in India is 6 per cent of the total population. Most of the increase in urban population is accommodated in slums. Mumbai and Calcutta have 40 per cent slum population, Delhi has 35 per cent, and Chennai 30 per cent. The average slum population in the mega-cities is 35 per cent. Slum population plays an important role in planning, because slums put a great strain on the general economics of maintenance of schemes.

Preferential treatment of mega-cities

The mega-cities enjoy definite advantages over other towns owing to their political, industrial, and commercial importance. This has

Mega-cities in India

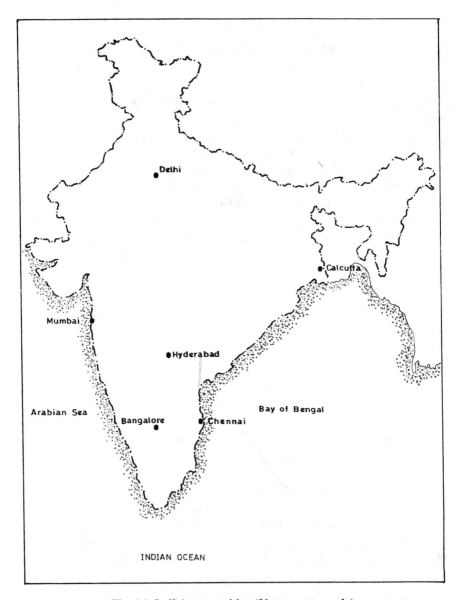

Fig. 4.1 **India's mega-cities (Note: not to scale)**

resulted in a great concentration of population there. Because of government's sensitivity towards the issues of mega-cities, these cities receive better attention from government and schemes of large magnitude and cost are approved more quickly, receiving priority in budgetary allocations.

Pursuant to recommendations by the National Commission of Urbanization, a centrally sponsored scheme for infrastructural development in mega-cities was introduced during 1993/94. The scheme covers Mumbai, Calcutta, Chennai, Bangalore, and Hyderabad, all of which had populations of more than 4 million in 1991. The state government and central government each take responsibility for 25 per cent of funding. The balance is raised through public finance institutions or the capital market. The Housing and Urban Development Corporation (HUDCO) has generally agreed to meet this component.

Problems of water supply in the four Indian mega-cities

Table 4.1 details the population and water supply in the four Indian mega-cities. There are several major problems common to all the cities.

Inaccurate population projections

Populations projected at the design stage are reached much earlier than anticipated. This necessitates immediate augmentation of water sources as well as the distribution networks. Hence, when framing any new scheme, stress has to be laid on population studies.

Rising population

Owing to the better availability of employment, education facilities, trade and commercial opportunities, etc. in the mega-cities, there is continuous migration of people from rural areas and smaller towns. Most of this influx results in the development of slums.

Rising demand for water

Owing to rising standards of living, changes in life style, and an increasing ability to pay, there is an increase in the consumption of water.

Inadequate distribution networks

Both rising population and rising demand lead to an inequitable distribution of water and pressures on an already inadequate system.

Table 4.1 **Water supply in the four mega-cities of India, 1997**

City	Population (million)			Slum population (%)	Water supply available (million litres/day)	Average per capita supply (litres/day)	Unaccounted for water (%)	Major sources (%)	
	1991	1997	2021[a]					Surface	Ground
Mumbai (Bombay)	9.90	11.00	16.00	40	2,600	236	40	100	–
Delhi	8.42	11.00	28.20	35	2,700	245	25	87	13
Calcutta	8.10	11.86	26.25	40	1,400	120	30	90	10
Chennai (Madras)	5.36	5.42	13.90	30	350	65	30	75	25

a. Projected increase of 4% per year.

The ever-growing number of new settlements puts excessive strain on the existing feeder mains, which are not designed for the increased density of population. Another serious problem is the reduction in carrying capacity of pipes owing to encrustation and damage with age.

Quality problems

Public water supply authorities are so busy trying to meet the quantity needs of urban people that the quality aspect is often given less attention than is warranted. Causes of poor-quality water include intermittent supply, which leads to back syphonage, low pressure, malfunctioning meters, lack of action on leak detection and repairs, poor supervision of the condition of pipes, and inadequate monitoring of residual disinfectant, which is a tracer of contamination.

Unaccounted for water

Unaccounted for water is very high in almost every city in India. The figures for the four mega-cities are as follows: Mumbai, 40 per cent; Delhi, 25 per cent; Calcutta, 30 per cent; Chennai, 30 per cent.

Local municipal bodies should lay maximum stress on water management, conservation, and the prevention, detection, and repair of leaks. Even if just visible leakages were stopped, the available supply would increase considerably, as would revenues. The conservation of water is an important part of supply management. To achieve this, a public awareness campaign, the participation of non-governmental organizations, and the training of women and children are essential.

Water supply in Mumbai

Mumbai, India's commercial capital, is the largest populated metropolis in the country and is one of the 10 largest mega-cities of the world. Its population is in excess of 11 million, and is forecast to reach 16 million by the year 2021.

Water sources

Mumbai's water supply is managed by the Municipal Corporation of Greater Mumbai (see fig. 4.2). Between 1860 and 1972, six sources were developed and these still supply water from rivers as well as

Mega-cities in India

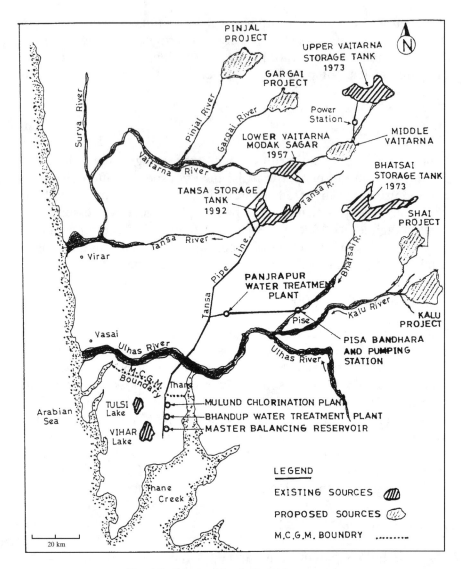

Fig. 4.2 **Greater Mumbai's water sources**

dams and lakes. Between 1981 and 1997, three more schemes, each with a capacity of 455 million litres a day (mld), were commissioned with World Bank assistance (under Bombay Water Supply and Sewerage Integrated Project I, II, and III). A fourth project for immediate implementation (Project III-A) has been put forward to the World Bank for assistance.

Present water demand is of the order of 3,530 mld, whereas supply is about 2,915 mld. The supply is unlikely to cope with demand even by the end of the first decade of the twenty-first century.

Of the supply, 80 per cent is for domestic use and 20 per cent is non-domestic. However, for revenue receipts the figures are reversed, i.e. 80 per cent of the revenue comes from non-domestic users and 20 per cent from domestic users.

The average minimum per capita supply in Mumbai varies as follows: in slums (from standpipes), 45 litres; in chawls (flats with shared toilet facilities), 90 litres; in flats and new buildings, 135 litres; in posh areas and five-star hotels, over 200 litres.

In addition to the piped water supply, there is a small amount of borewell water (1 per cent of the total supply). A subsidy is given by the Municipal Corporation for constructing borewells. Supply in Mumbai is intermittent – up to four hours a day.

Problems in water management

Both the drainage system in Mumbai and also the entire old water distribution network are in bad condition. Consequently, every source of leakage is a source of contamination. In order to conserve water and also to check contamination, leak-proofing of cracks and joints is being done. It is planned to line 350 km of pipes with mortar by 2002, of which about 40 km has been completed by working during non-supply hours. Some of the old pipes have developed encrustation and weed growth, and some are even blocked by stones, which has reduced the pressure and the supply. The casting of very old pipes is good, but new pipes deteriorate fast. As a result, the working pressure falls as low as 1.5 m of water head even though the designed pressure at the tail end is 10 psi (7 m). It is proposed to replace 130 km of pipeline in the near future.

There are maintenance problems too, for example in repairing leaks, due to the proximity of telephone and power cables, laid in the same trenches as the old pipelines. Moreover, the old concrete roads are not provided with ducts, so breaks or leaks under such roads necessitate the breaking up of the road, leading to long repair times. Congestion in side roads makes the diversion of feeder mains impossible.

Old two-storey houses are being reconstructed into multi-storey apartment buildings, causing an almost three-fold increase in demand.

Housing in the largest slum in Asia, Dharavi in Mumbai, is improving, thus creating increased demand.

Raw water is brought from neighbouring districts, whose residents expect water from these mains, and work on new pipelines has often been stopped by them, resulting in delays in completing augmentation schemes.

Suggestions to improve the situation

Supply management

Mumbai has had better availability of water compared with other cities. However, development of Mumbai's water supply is mostly demand driven and no efforts have been made to manage demand as such. Since future sources are going to be much more costly, proper management of the system to satisfy demand is essential.

Although groundwater potential is only 3 per cent of expected demand in the year 2021, it has to be harnessed and used for non-domestic purposes. In addition, other sources need to be found to supplement the available supply.

Alternative sources

REUSE AND RECYCLING OF WASTEWATER. The reuse of wastewater was started in 1960 in textile mills and there is still vast scope for it in industry. Industrial water demand in Mumbai is not likely to increase substantially (present consumption is 205 mld and by 2021 it would be 700 mld). Even so, a 15 per cent reduction in freshwater intake by industries would mean a reduction in demand of 100 mld by 2021.

REUSE OF DOMESTIC SEWAGE AFTER PROPER TREATMENT. There is great scope for using treated sewage as non-potable water for flushing toilets, car washing, fire-fighting, cooling in commercial buildings, gardening irrigation, and the cleaning of railway wagons. In order to promote such practices, some incentives could be introduced. An appropriate target for the production of such non-potable water would be 2 per cent of supply in 2021.

SEAWATER DESALINATION. Large seawater desalination plants have not so far been constructed in India. It is an expensive process, in terms of both capital as well as operating and maintenance (O&M) expenditure (costing Rs. 25–30 per m^3). However, medium-sized reverse osmosis plants, treating brackish water, are in operation.

There is scope for desalination in Mumbai along the sea coast, in isolated and high-value establishments such as hotels, defence establishments, oil refineries, fertilizer factories, and atomic energy establishments. By 2021, 4 per cent of total demand in Mumbai would be of this scattered type. Large industries could adopt this method and save potable water. Hence, more detailed technical and financial analysis of potential desalination methods will have to be undertaken to provide proper guidance for possible installation in the near future.

According to a recent study, a desalination plant with a power house producing 1 million gallons per day would cost Rs. 200 million. The cost of the desalinated water would be Rs. 45/m^3, against the present rate for potable water of Rs. 32/m^3; this increase should be acceptable in a time of need and also as a social obligation.

Artificial rain making
Several "warm cloud seeding" experiments have been tried but the technique is still in the research stage.

Evaporation control
Although evaporation control in reservoirs using evaporation retardant chemicals has been tried in the past, further studies are needed, because wave action is a major obstacle to this method. The integrated operation of reservoirs to reduce the total exposed water surface area by depleting the reservoirs with a large exposed water area first would be very beneficial in arresting evaporation. This is important in view of the fact that there will be 10 reservoirs serving as sources by 2021.

Leak detection
The prevention of leaks would save 25 per cent of the water produced and also avoid contamination of water. Hence, a reduction by at least 15 per cent should be the target. The optimal methods for achieving this need to be studied.

Recycling of wastewater from water treatment plants
The wastewater from settling tanks and backwashing filters needs to be recycled and returned to the system. This would add 2 per cent to the supply. Not all plants have this facility.

All these conservation methods are likely to save about 1,046 million litres of water, which would reduce the cost of future schemes by 40 per cent.

Demand management

Sources are earmarked up to 2021, but by the middle of the twenty-first century, if demand continues to grow at the same rate, it will be difficult to find sources other than seawater. Hence it is not prudent to plan water supply schemes on the basis of exponential growth. Rather, demand should be controlled through special deconcentration measures. This is essential in view of the increasing cost of additional supplies, the difficulty of repairing and replacing existing systems, and a shortage of land for additional components. The government has already started moving in this direction by shifting growth from Mumbai to Navi Mumbai.

Financial management for a self-supporting system

The rate for the domestic water supply in Mumbai is the lowest in India. Until recently, it was Rs. $0.60/m^3$, compared with a production cost of Rs. $2.75/m^3$ (the average cost of production of water in India is Rs. $6.40/m^3$). Domestic water is highly subsidized by the high rates for the non-domestic and commercial water supply – 80 per cent of revenue comes from the 20 per cent of non-domestic and commercial supply. However, the calculations for deriving this rate were not realistic because they included only the O&M costs as expenditure, without accounting for debt servicing charges, etc.

Owing to rising costs, the actual marginal cost of water in real financial terms will be much higher for future schemes; subsequent tariff adjustments should therefore be based on the principle of marginal cost. A proper study and framing of tariffs is therefore required such that the revenue generated leaves a surplus for future capital works.

The current tariffs
Mumbai municipal corporation has recently raised water rates substantially for the first time; this has automatically increased the sewerage cess, which is 50 per cent of the water bill. The increase in revenue is to be utilized for undertaking O&M works more rigorously and for capital expenditure for future schemes.

The new domestic rate is also differential, as follows: Rs. 1.00/m^3 for houses in slums; Rs. 1.50/m^3 in old chawls; Rs. 2.00/m^3 in private housing associations; Rs. 2.75/m^3 for multi-storey flats and bungalows.

Collection of water charges

The recovery of water charges has been highly inefficient. There are arrears of about Rs. 2,500 million. The defaulters are mainly state government offices and central government establishments. The government therefore needs to give directives to its departments to pay both their arrears and also their current bills promptly.

Automation

In the past, O&M has always been labour intensive. It is only now that the municipal corporation is thinking in the direction of automation of instrumentation and has invited tenders for instruments to measure flow and pressure in pipelines, chlorine sensors, etc.

Planning for the future

Table 4.2 shows the dams that are planned for the future. By 2021, however, supply will outstrip demand. It is likely that seawater desalination, saving through conservation, and using alternative sources will be the only way to augment water supplies to meet future demand.

Water supply in Delhi

Delhi, being the capital of India, is an important city. The capital was shifted here from Calcutta in 1912. The city has grown very fast since independence, and is still expanding. The influx of population from

Table 4.2 **Proposed dams to augment Mumbai's water supply**

	Daily supply (mld)	Year
Vaitarna Basin		
Middle Vaitarna	477	2003
Gargai	455	2009
Pinjal	865	2021
Ulhas River Basin		
Kalu	595	2007
Shai	1,067	2013

adjoining areas and states continues. Delhi has the highest population density amongst Indian cities (12,953/km^2 in 1991). In 1941, the population was less than 1 million, today it is 11 million. The growth of population during the decade 1941 to 1951 was 106 per cent and since then it has averaged 56 per cent per decade. Per capita water supply varies widely from low in slums to very high in high-class areas of New Delhi. The Delhi Water Supply and Sewage Disposal Undertaking is in charge of water supply and sewage disposal in Delhi, New Delhi, the cantonment, and the villages.

The present water supply system, peculiarities of the system, management problems, and planning for future are briefly described below.

Water sources

Surface water

Although Delhi is situated on the banks of the River Yamuna, it enjoys no riparian rights. The entire flow of the river is impounded 250 km upstream of the Vazirabad water treatment plant for diversion to Hariyana and Uttar Pradesh states. Delhi is thus dependent on neighbouring states for surface water. Its water comes from four sources (see fig. 4.3):

(a) From December to June water is indented from the Bhakra Bias Management System for the Haiderpur Water Treatment Plant, through the Western Yamuna Canal.
(b) Water is released 80 km upstream on the River Yamuna for the Vazirabad Water Treatment Plant.
(c) Surface flow during the rainy season and regenerated water are available from the River Yamuna for Chandrawal Water Works I and II.
(d) Ganga water for the Bhagirathi plant at Shahadara is supplied from the Tehri dam storage through the Upper Ganga Canal, 24 km upstream of the treatment plant site, and conveyed to the treatment plant through 2,800 mm diameter reinforced cement concrete conduits.

Groundwater

Hydrogeologically, Delhi has limited groundwater of the required quality and there is a widespread chemical quality problem (in particular, a high fluoride content results in the incurable disease of skeleton fluorosis).

Mega-cities in India

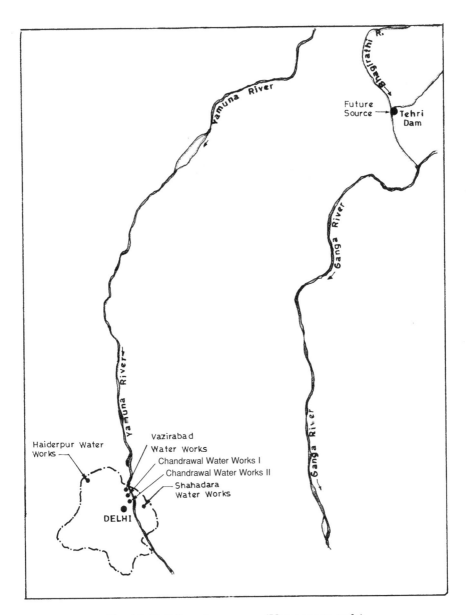

Fig. 4.3 **Delhi's water sources (Note: not to scale)**

Although raw surface water supplies are comparatively uncertain, the availability of drinking water in Delhi is much better than in any other mega-city (in spite of the fact that water consumption for commercial and industrial uses has increased):

(a) the hours of supply are longer;
(b) per capita availability is greater, having increased from 190 litres per day in 1971 to about 245 litres per day today;
(c) the unit production cost is low and the average domestic tariff is also quite low (Rs. 0.35–0.70/m^3); in addition, almost 900 mld are supplied free. Domestic water is subsidized by high non-domestic and commercial rates (Rs. 3–5/m^3). Operation and maintenance costs none the less exceed revenue.

In addition to the piped water supply, water is obtained from various other sources such as tubewells, deep bore handpumps, ranney wells (radial wells sunk in the bed of a river), and supply by tanker, where needed.

The water supply system in Delhi is very large. There are nearly 12,000 km of pumping, peripheral, and distribution mains and 384 booster pumping stations. The number of water connections is 1.2 million. Water is supplied to nearly 90 per cent of the population, covering 567 unauthorized and regularized settlements, 1,080 squatter settlements, 44 resettlement colonies, 413 Harijan *bastis* (clusters of small houses), 222 rural villages, and 226 urban villages. The length of distribution mains increased from 3,742 to 7,400 km between 1985 and 1996. The strength of staff on operation and maintenance is approximately 11,000.

Problems in water management

All water treatment plants in Delhi are on one side of the city whereas consumption is on the other side. This results in uneven pressure in the distribution system at tail ends. Underground tanks and booster pumping stations have thus been constructed. The aim is not to supply water throughout the day but to achieve an adequate supply during peak hours. During breakdowns in the system, filling points for supply by tankers are essential to restore the water supply until repairs are completed.

Water pollution is a higher priority because there are epidemics of waterborne diseases every year. Steps have been taken to prevent pollution of the River Yamuna by diverting polluting streams to downstream of intake points. However, the river downstream is still getting polluted from the discharge of sullage through Nallas and other outlets, and industrial pollution still exists upstream of intakes. Groundwater is also polluted and contains fluorides. When tubewells were constructed they were not tested for quality. Hence, they are no use for domestic purposes and residents have been made aware of this.

The condition of old pipes is bad and, because water and sewage pipes are close together, there is cross-contamination. To remedy this, broken pipelines in the distribution system are being replaced in a phased manner. There is, however, an effective surveillance system to test the quality of water at all points, from the raw water stage to the consumer. Nearly 450,000 service connections require replacement and about 71,000 connections were replaced in 1996. Consumers are being encouraged to replace pipes to avoid contamination.

Problems causing working constraints

In addition to the detection and repair of leaks, which are being undertaken by an independent unit, overhaul of the distribution system is essential so that the water that is saved is added to the available supply. Similarly, huge wastage of water through public standpipes needs to be monitored and curbed, and illegal tapping prevented.

Even so, rising standards of living mean that demand will continue to outstrip supply.

The management of the system is also affected by a floating population of tourists and visitors to the capital of the order of 300,000 to 400,000. Moreover, the regular influx of population has increased during recent decades. It is therefore essential to monitor future projects to augment supply.

An increasing trend towards commercialization and densification, as well as continuing unauthorized construction, has increased demand for fire-fighting.

Scope for improving the water supply and water management

The possibility of utilizing more groundwater in a scientific manner is being explored seriously, particularly in the outlying areas of Delhi. This water could be treated and used for drinking. What was uneconomic a decade ago will become a necessity in future.

Also under active consideration is the possibility of exchanging the stormwater from drains as well as treated effluent from the sewage treatment plants of Delhi with the river water being used for irrigation by the neighbouring states of Haryana and Uttar Pradesh. This river water could be used in Delhi for drinking purposes.

The use of filtered water for parks, gardens, and kitchen gardens will have to be banned totally, and should be substituted by borewell water, recycled water, or treated effluent from sewage treatment plants.

A rationalization of supply hours is also needed in order to ensure that water is used for essential requirements only.

Water management thus deserves to be a focal issue in the overall framework of urban development and management of urban services in Delhi. There have been a few success stories in Delhi where consumers themselves have mobilized opinion and action to restrict the misuse of this scarce resource. In south Delhi, some apartments have developed a dual system of water supply and use. Potable water is stored in one elevated tank for the entire building for drinking purposes and borewell water is supplied for other uses through a separate tank and system. This not only ensures a regular supply of potable water but also encourages good demand management. A public awareness campaign is needed to educate people about such methods and about conservation.

Water conservation

Various methods have been adopted to conserve water.

Evaporation control
Evaporation is controlled by carrying raw water from the Upper Ganga Canal to the Bhagirathi water treatment plant in a closed conduit. In addition, the Yamuna Canal raw water carrier system to the Haider plant is lined.

Recirculation of wastewater and recycling of waste
Wastewater from clarifiers and filter wash water are already being recycled at the Bhagirathi water treatment plant. Adoption of the system is being studied at other plants. Recycling of wastewater for industrial use has also been started. In addition, 270 mld of sewage effluent is being used for irrigation in the Union Territory of Delhi.

Prevention of leaks, wastage, misuse, and theft of water
Unaccounted for water, which was 40 per cent in 1992, has decreased to 25 per cent as a result of various measures taken in the past. Vigorous efforts need to be made to install a remote sensing and monitoring system to reduce the detection and response time for leaks, bursts, etc.

Mega-cities in India

Table 4.3 **Planned dams to augment Delhi's water supply**

	Potential (million m^3)	Delhi's share (million m^3)
Tehri Dam	1,858	268
Renuka Dam (in Himachal Pradesh)	460	460
Kishau Dam (in Uttar Pradesh)	1,300	615

A special drive has been launched to remove on-line boosters illegally installed by consumers – 17,826 boosters had been removed by March 1996.

Planning for the future

Delhi's water supply capacity from all sources is about 2,700 mld (of which 250 mld is from wells and tubewells).

Work to increase the supply is currently in progress, consisting of the construction of water treatment plants at Nagloi (182 mld capacity), treating water from the Western Yamuna Canal, and at Bawna (91 mld capacity), and of five ranney wells. To cater for the prospective population, three dams are planned (table 4.3).

Water supply in Calcutta

Calcutta is an important coastal town in the east. Greater Calcutta includes Howrah, which is the railway terminus.

Water sources

The sources of Greater Calcutta's drinking water supply are shown in table 4.4 and figure 4.4. Within the Calcutta Municipal Corporation limits, the construction and maintenance of water supply schemes are undertaken by the Calcutta Municipal Corporation (CMC) and outside the city limits by the Calcutta Metropolitan Development Authority (CMDA). The Calcutta Metropolitan Water Supply and Sewerage Authority is responsible for the expansion of waterworks.

The population of Greater Calcutta in 1997 was about 11.86 million, with a floating population of 2 million. The slum population was about 40 per cent. The average daily water supply per capita was 120 litres.

Table 4.4 **Greater Calcutta's sources of drinking water**

Source	Volume	
	Million gallons/day	Million litres/day
Calcutta		
Surface water (River Ganges):		
Palta waterworks (30 km away)	160	730
Garden Reach waterworks (16 km away)	60	270
Baranagar Kamarhati waterworks (CMDA) for BK area of Greater Calcutta	30	136
Total	250	1,136
Groundwater:		
Various zonal tubewells	20	90
Private pumps	10	45
Total	30	135
Howrah		
Surface water (River Ganges):		
Padhmapukur waterworks	40	180
Serampore waterworks	20	90
Total	60	270
Total	340	1,541

Distribution system

Water is supplied at high pressure (5 kg/cm^2) between 6 a.m. and 9 a.m. and 3.30 p.m. and 6.30 p.m. and at low pressure between 11.00 a.m. and 12.00 noon and 8.00 p.m. and 8.30 p.m. Residual heads are generally 0.60 m.

Water tariffs

Domestic
A large proportion of the filtered water supply in Calcutta is not directly charged because the water supply charges are included in the property tax. In Calcutta Municipal area, water connections up to 15 mm diameter are not charged, 20 mm diameter connections are

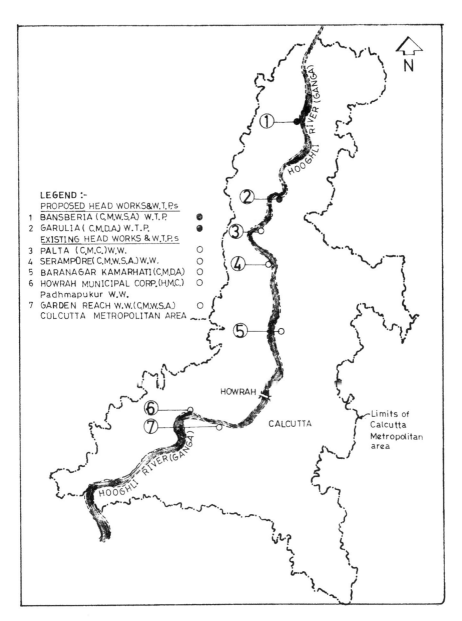

Fig. 4.4 **Tapping points on the River Ganga for Calcutta Metropolitan Area (Note: not to scale)**

charged at Rs. 480 per year, and 25 mm connections at Rs. 780 per year. Bulk water supplies are often metered and charged at Rs. $3/m^3$.

Commercial and industrial
For filtered water, a 6 mm connection costs Rs. 240 per month and a 25 mm connection Rs. 2,200 per month.

Some unfiltered water is also supplied (by disinfecting raw water from the River Ganga) to industries at nominal cost, varying from Rs. 50–75 per month for a 12 mm diameter connection to Rs. 335–505 per month for a 25 mm diameter connection.

Bulk water supplies to industries are charged at Rs. $15/m^3$.

Management problems

Calcutta faces problems of overcrowding and narrow lanes in the Howrah area, making O&M repairs difficult.

There is arsenic pollution in the groundwater, which is being drawn from greater depths because of the falling water table. As a result, deep tubewells are being closed.

About 50 per cent of the water supply is not directly charged, which creates financial problems in terms of O&M, even though domestic water is cross-subsidized.

For any new scheme, funding is a problem because government grants cover only 25 per cent of the cost. Funding from other financial institutions is required, which is a slow process. Raising funds by floating public bonds is being tried now.

Unaccounted for water

The percentage of unaccounted for water is about 30 per cent.

Planning for the future

New capacity
Future population and demand for water are shown in table 4.5. To meet this demand, it is proposed to add 40 million gallons/day (mgd) at the Palta plant, 60 mgd at the Garden Reach plant, and 40 mgd at the Howrah waterworks by the year 2000, and another 40 mgd each at the Palta and Garden Reach waterworks by the year 2021.

A new project, estimated to cost Rs. 5,000 million, is at the conceptual stage. This envisages procuring the entire water supply from

Table 4.5 **Projected population and demand for water in Calcutta, 2000 and 2015**

Year	Population (million)	Demand (mld)	
		Greater Calcutta	Calcutta Corporation's share
2000	15.50	2,034	675
2015	21.40	2,565	696

treatment works by gravity, thus making considerable savings in power consumption. The gravitational flow would also ease operational problems.

Rehabilitation of the distribution network
The pipes used in the distribution network are very old. Their carrying capacity has fallen, making it difficult to distribute water equitably at adequate pressure.

A rehabilitation plan for large-diameter pipes (750–1,800 mm diameter) is in hand, for which a separate allocation of Rs. 500 million was made during the 1997/98 financial year. Fibreglass reinforced pipes are being pushed inside the old cast-iron pipes in the rehabilitation process.

Water supply in Chennai

Chennai, the capital of Tamil Nadu state, is situated on the coast of the Bay of Bengal. Though the city receives an appreciable amount of rainfall (1,000–1,200 mm), drinking water is the most serious problem every year. Although there are five rivers in and around, the dependable supply from these rivers is insufficient to meet the needs of the city. Chennai has a population of 5.42 million and the total water supply is of the order of 300–350 mld. Thus, the per capita supply is only 65 litres, which is too low. Rationing of drinking water is required, including alternate-day supply when necessary. In the past, water was transported by train.

Water sources

The present water supply comprises about 75 per cent surface water and 25 per cent groundwater.

Surface water

The present source of surface water is natural rainwater stored in three interlinked lakes: Pundi, Cholawaram, and Red Hills (see fig. 4.5). However, all three lakes are shallow and small in capacity. The total water obtained from this system is approximately 250 mld.

The catchment area is an agricultural region, so that in years of low rainfall the inflow to the reservoirs reduces considerably because irrigation tanks, lakes, and upstream diversion have first call on the available flow.

Groundwater

About 50 mld is extracted from well fields. Although groundwater makes an appreciable contribution (equivalent to industrial requirements in the area), there are certain problems with aquifers that are shallow and have limited resources. As a result of great exploitation of groundwater, some well fields have experienced seawater intrusion. In consequence, water from almost 40 per cent of the wells is not of potable quality. In addition, a major recharge area in the town is now covered by urban settlements, leading to a decrease in groundwater because of low groundwater recharge. There needs to be a check on such settlements.

Projects to augment the supply for Chennai

Because Chennai's water supply has been far from satisfactory, several schemes have been formulated from time to time in the past to augment water sources. However, there have been a lot of political problems and the proposals were cancelled.

Krishna Water Supply Project

The first augmentation scheme, the Krishna Water Supply Project, was conceived in 1983. It planned to draw water from the River Krishna in the state of Andhra Pradesh, 400 km from Chennai. However, this scheme did not ultimately materialize.

Viranam Water Supply Scheme

A project to bring 200 mld from Viranam lake, 230 km away, planned originally some 15 years ago, was abandoned mid-way through its execution. Attempts were made by successive governments to renew this scheme by planning it all over again and linking it up with the World Bank. But, owing to uncertainties about the availability of the

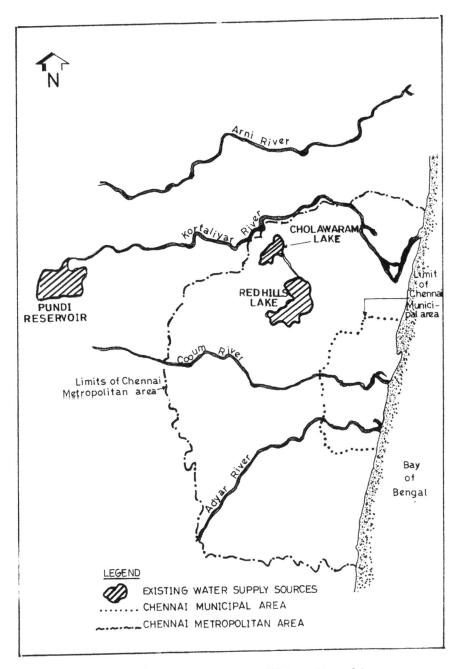

Fig. 4.5 **Chennai's water sources (Note: not to scale)**

required volume of water, the scheme has now been downsized and redesigned to serve towns in adjoining districts. Thus this source is not available to augment the city's water supply.

A scheme for the future
Another source has been identified, and a scheme is being carried out to draw water from the River Krishna through the Nagarjun Sagar dam, 900 km away. It is in two stages: the first stage of 400 mld is nearing completion; in the second stage, another 500 mld will be made available.

By the time the second stage is completed, it is thought that another 20 local bodies around Chennai with a population of about 3 million will have to be served from this source, by which time the total available supply will be 1,200 mld. This could cater for a population of 10–11 million in the future. A large unserved area around Chennai will have also to be served from this water supply scheme, which would reduce the per capita availability.

Water management in Chennai

The Madras Metro Water Supply and Sewage Board (MMWSSB) has been very aware of the problem of the scarcity of water in Chennai, and it has also started taking firm steps to conserve existing potable water and to use it economically and specifically for drinking purposes. One strategy is that the government does not sanction any industrial project unless the industry makes its own water supply arrangements or uses recycled wastewater, duly treated for the industrial consumption. To cite one case, Madras Fertilisers Ltd. was provided with 18 mld raw water by the MMWSSB from aquifers located 10 km from the factory site. During severe droughts in 1983 and 1987, the MMWSSB imposed a total ban on the water supply to this industry; this led to a long shutdown at this plant. The industry sought to replace its requirement of 12 mld for cooling with an alternative source and evaluated the possibility of either utilizing seawater for an indirect cooling process or supplementing the existing water supply with city sullage requiring tertiary treatment and demineralization prior to use as a feedstock for the existing factory water supply system. The second option was chosen and a plant using reverse osmosis to convert 15 mld of effluent into 12 mld of usable industrial-grade water suitable for their cooling towers was constructed. It was commissioned in 1993. The freeing of 12 mld of potable water for the

Chennai water supply system helps to ease the social pressures of an ever-increasing population.

The MMWSSB accepted the principle of providing industries with water obtained from purified sewage effluent. Japan's Overseas Economic Cooperation Fund (OECF) has agreed to fund a 100 mld treatment plant to treat a portion of the city's sewage, thus creating another industrial water resource and in effect augmenting the city's water supply.

Other measures for water conservation

In the past, evaporation control by spreading chemicals on the lakes has been tried and a saving of 25 per cent was observed. This measure could be adopted in the future when needed.

In the water treatment plant, filter backwash is reused. In addition, a desalination plant is under construction.

Under the Madras Metropolitan Groundwater Act, no new site plans can be sanctioned without provision for rainwater harvesting. There is also a restriction on the capacity of pumps for drawing groundwater. As a result, there has been a perceptible rise in the groundwater table.

Apart from these measures, leak detection and prevention are being pursued in a big way. Thus, in the future, the situation is expected to improve.

Conclusions

Several strategies need to be adopted to achieve a sustainable water supply in India's mega-cities:
- A national water policy to decide priorities in water allocation and for the quick settlement of inter-state disputes should be put into operation. A groundwater act, to halt over-exploitation, is needed. To prevent the pollution of surface water as well as groundwater, polluters should be booked and legal acts enforced.
- The seventy-fourth amendment to the Indian Constitution, which delegates financial powers to local bodies, should be implemented.
- When framing schemes, population forecasts need to be made scientifically, since this is the basis for design.
- Growth in the cities has to be controlled because it leads to the development of slums.

- Unaccounted for water is very high; hence a systematic leak detection and leak prevention programme, as a part of a regular O&M plan, is a first priority.
- Water conservation has to be given due weight. Water should be appropriately priced because low rates tend to increase wasteful use of water. However, there should be subsidy to the poor. Simple in-house conservation measures such as self-closing taps and small-capacity toilet cisterns should be taken.
- Water tariffs should be calculated on a marginal cost basis, and yearly increases of 10–15 per cent made. Indirect taxation should also be tried, if necessary. Water supply and sewerage should be charged for separately and there should be an effective recovery mechanism.
- Water quality and surveillance should be accorded the same priority as the quantity of supply. Checks on residual disinfectants and proper testing should be insisted on. Old pipes and connections are the source of contamination, so there should be a plan to replace these in stages.
- To plan for the future, a 10-yearly review of supply and demand and long-term planning for 30–50 years should be undertaken. Integrated water supply and sewerage schemes should be prepared.
- Much of the cost of new schemes with distant sources could be saved by meeting some demand through alternative sources such as: (a) recycling of sullage and wastewater for non-potable use; (b) using secondary treated sewage for industrial and agricultural purposes; (c) exploring groundwater potential; (d) seawater desalination in high-value establishments in coastal cities.
- Metering is desirable provided the supply is more or less continuous, the quality of meters is good, and servicing back-up is available.
- Good organizational linkages and interdepartmental coordination and cooperation are the key to avoiding delays in implementation.
- Institution-strengthening and capacity-building through training are vital.
- Political stability, will, and support are required for long-term planning. Planning should not be limited to the tenure of elected representatives.
- Resource mobilization and financial planning are central to any project. Dependence on government funds has to be reduced and innovative ways of financing should be devised; these include institutional finance and external funding.

- Last but not least is the very important place and role of engineers in the city's water management. They are better equipped to understand the problems involved than generalists and bureaucrats are. However, proper publicity by engineers about their work is necessary. They need to establish a rapport with the public and educate it.

Note

1. Throughout this chapter, the local name of Mumbai is used for Bombay, and the name Chennai is used for Madras.

5

Water supply and distribution in the metropolitan area of Mexico City

Cecilia Tortajada-Quiroz

Introduction

Mexico is a country with an area of 2 million km² and a population of more than 92 million. The mean annual rainfall is 780 mm (representing a volume of 1,522 km³), the average annual runoff is 410 km³/year, and the renewable annual groundwater is estimated at 55 km³. The average annual per capita water availability is about 5,000 m³, which is twice the world average. However, water is scarce in the north of the country and abundant in the south. In fact, 79 per cent of the natural groundwater recharge occurs in the south-eastern region of the country (SEMARNAP/CNA, 1996). Mexico's per capita water availability thus varies regionally, depending on both the mean annual rainfall and population concentrations: in regions with less water but high population, availability ranges between 211 and 1,478 m³/year; in regions with more water and less population, availability varies between 14,445 and 33,285 m³/year (CNA, 1994).

In 1995, the total water abstracted in Mexico for all purposes was estimated at about 300 km³. Of this, 26.5 per cent was allocated to produce hydroelectric energy and 73.5 per cent for other purposes: 61.2 per cent for agriculture, 8.5 per cent for domestic use, 2.5 per cent for industry, and 1.3 per cent for aquaculture (fig. 5.1). It is

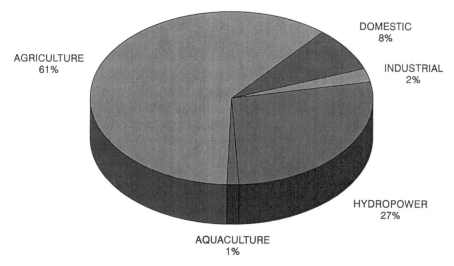

Fig. 5.1 **Water use by sector in Mexico, 1995 (Source: SEMARNAP/CNA, 1996)**

estimated that, by the year 2000, 142 km^3/year will be required to produce hydroelectric energy and 2.89 km^3/year for thermal power (mainly cooled by seawater) (SEMARNAP/CNA, 1996).

Current annual average water abstraction represents about 43 per cent of total annual renewable water. Viewed nationally, this may give a false sense of water abundance since it does not indicate the problems of scarcity and contamination that can now be observed in most of the basins and aquifers. Both water scarcity and water contamination have already contributed to the development of a series of conflicts between the various uses and different users in many parts of the country (Naranjo and Biswas, 1997; CNA, 1994).

The total annual wastewater generated in Mexico is estimated at 7.3 km^3 (or 231 m^3/sec). Even though infrastructures exist to treat 1.4 km^3/year, only 0.53 km^3/year are actually treated. Thus, nearly 93 per cent of the wastewater is discharged without any treatment whatsoever (SEMARNAP/CNA, 1996). The 1994 studies by Mexico's National Water Commission (Comisión Nacional del Agua, CNA) on water quality in 218 basins in the country (representing 77 per cent of the territory, 93 per cent of the population, 72 per cent of the industrial sector, and 98 per cent of irrigated areas) indicated that most of the basins are at present contaminated with organic, industrial, and/ or agrochemical wastes (CNA, 1994).

The number of people being supplied with water has increased in recent years (fig. 5.2), but not all of this water is treated (fig. 5.3).

The metropolitan area of Mexico City

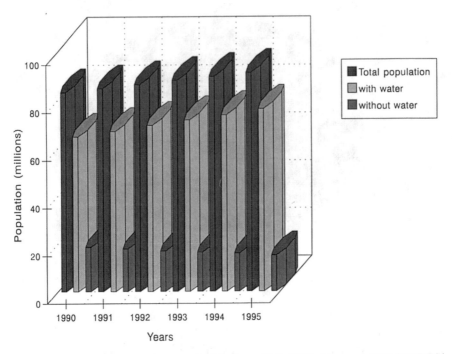

Fig. 5.2 **The availability of water in Mexico, 1990–1995 (Source: CNA/UPRPS/ National Information System – CNA, 1997a)**

According to the CNA (1997a), out of a total population of 91.6 million, 15.1 million people did not have access to clean drinking water and 30.6 million did not have access to sanitation facilities in 1995. The problem is most serious in the rural areas, where 48 per cent of the population do not receive safe drinking water and 79 per cent do not have access to sanitation services (SEMARNAP/CNA, 1996).

Current estimates indicate that, by 2000, total demand for drinking water in Mexico will increase to 9.4 km^3/year (299 m^3/sec), and 7.7 km^3/year (244 m^3/sec) of wastewater will be generated (SEMARNAP/CNA, 1996). Already, however, Mexico faces serious problems of water availability because of the mismatch between the centres of water demand and of available water. The 75 per cent of the territory in which most of the large cities, the industrial sector, and irrigated land are concentrated accounts for only about one-third of the water available in the country.

The problems related to water management have become very complex. The increasing population, the very serious problems related to lack of sanitation and clean water, as well as permanently

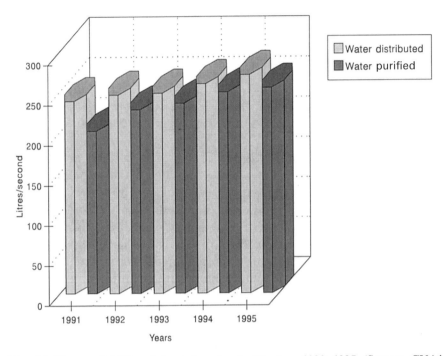

Fig. 5.3 **The availability of treated water in Mexico, 1990–1995 (Source: CNA/ UPRPS/Department of Clean Water – CNA, 1997a)**

high investments in infrastructure without achieving definite solutions are forcing the government to search for different approaches other than purely engineering and supply-oriented ones. The institutions concerned are realizing that successful water resources management requires a long-term planning process from the technical, economic, political, social, and environmental points of view.

A good example of unsustainability from all the different points of view is Mexico City, where both water supply and wastewater treatment have become serious problems. So far, no long-term sustainable strategies have been proposed by the government, either federal or state, and, unless programmes on demand management for the city are seriously considered and implemented, it will be only a matter of time before more acute conflicts arise between the different users of clean water.

Mexico City: A case study

Mexico City is the capital city of Mexico. It is located in the Federal District (Distrito Federal, D.F.), in the south-western part of the

Valley of Mexico, surrounded by mountains reaching altitudes of over 5,000 metres above mean sea level (msl). At the beginning of the twentieth century, the city was still in the north-central area of the D.F. As a result of increased urbanization, it now covers the whole surface area of the D.F. The federal government and most of Mexico's industries, educational and employment facilities, and cultural centres are concentrated in this area (National Research Council et al., 1995).

Mexico City represents 0.1 per cent of the surface area of the country, and accounts for nearly 10 per cent of its population (8.5 million inhabitants) (INEGI, 1996). It is located in a high, naturally closed basin, at 2,240 msl, and most of the urban area is located in the flat and lowest levels of the basin. Historically, the city has faced severe problems related to water scarcity, which have become more acute owing to the continuous increase in population and the contamination of surface water and groundwater within and around the city.

In order to meet this escalating water demand, the government has focused mainly on supply management and engineering solutions, which have resulted in investments of millions of dollars and the construction of major infrastructural projects for the inter-basin transfer of surface water and groundwater. Long-term economic, social, and environmental strategies still have to be developed to achieve a more sustainable development of the basin and to improve the life styles of its millions of inhabitants.

In the Valley of Mexico, annual precipitation is about 746 mm (226 m^3/sec) (1980–1985) and evapotranspiration is about 75 per cent (Birkle et al., 1996). The city's water supply depends mainly on local groundwater sources and on the transfer of surface water from increasingly distant basins. In order to meet part of the water needs of the population living in Mexico City, a total volume of 1,199 km^3/day is abstracted from 347 wells (1,142 km^3/day) and 62 springs (57 km^3/day) (INEGI, 1996). The second main water source originates from the Lerma-Balsas and Cutzamala river systems.

The water is distributed to users through a primary network of 870 km of pipelines and a secondary network of 10,600 km. The water supply system comprises 14 dams with a total storage capacity of 2,332,700 m^3 and a catchment area of 170 km^2 (INEGI, 1996). There are also 243 storage and regulatory tanks with a total capacity of 1,500,000 m^3. Water is distributed to people living in the highest areas of the city by means of 227 pumps (UNAM, 1997).

Mexico City currently has 1,470 km of primary sewerage network and 9,900 km of secondary network. The volume of wastewater that is discharged is 2,349,116 km^3/year (INEGI, 1996), of which no more than 7 per cent is treated. There are 22 treatment plants in Mexico City (18 secondary level and four tertiary), with an installed capacity of 6,810 litres/sec (CNA, personal communication). The reports, however, do not specify whether all the treatment plants are currently functional or the extent to which their capacities are used.

The quality of life of the population living in the Mexico City area has decreased dramatically over recent years, primarily owing to the high density of population (ranging from 300/km^2 up to 10,000/km^2) and extensive air and water pollution. The increased urbanization and high population growth within the city have resulted in the designation of an area known as Mexico City Metropolitan Zone (Zona Metropolitana de la Ciudad de México, ZMCM). Thus, Mexico City can no longer be considered to be an independent unit for water planning and management. The ZMCM includes Mexico City and part or all of the 17 municipalities of the State of Mexico, which adjoins the D.F. to the north, east, and west.

Mexico City Metropolitan Zone

Water demand

The ZMCM is one of the most rapidly growing urban centres in Mexico. It has a surface area of 3,773 km^2 and more than 20 million inhabitants. The total population is not known reliably because of the very high rate of immigration as well as numerous illegal settlements (National Research Council et al., 1995).

The State of Mexico is the most populated area in Mexico (more than 12 million inhabitants according to 1995 reports), followed by Mexico City with nearly 8.5 million people (INEGI, 1996). The State of Mexico has an annual growth rate of 3.75 per cent (compared to the national growth rate of 2.43 per cent), an average population density of 545/km^2, and some 3,000 industries (CNA, 1997b; CONAPO, 1997).

The per capita water supply is 364 litres/person/day in Mexico City and 230 litres/person/day in the State of Mexico, which represents average daily consumption in the ZMCM of 297 litres/person (table 5.1; National Research Council et al., 1995). However, the actual amount received by individuals is significantly less, because the sup-

Table 5.1 **Characteristics of the Mexico City Metropolitan Zone and use of water supplied to Mexico City and the State of Mexico**

	Mexico City	State of Mexico
Total area of the ZMCM (km^2)	1,504	2,269
Area served by the common water distribution and wastewater disposal systems (km^2)	667	620
Population (million)	8.5	12
Daily per capita water supply (litres)	364	230
Water use by category (%)		
Domestic	67	80
Industrial	17	17
Commercial and urban services	16	3

Sources: Departamento del Distrito Federal, 1992b; Comisión Estatal de Aguas y Saneamiento, 1993; INEGI 1991a. In National Research Council et al. (1995); INEGI (1996).

ply also serves industries and services and leaks account for more than 30 per cent (Arreguín-Cortés, 1994); moreover, there are differences in distribution to the various areas of the ZMCM (Casasús, 1994; CNA, 1997b).

Total water consumption in the ZMCM is about 62 m^3/sec for domestic and commercial uses, as well as for services – 35 m^3/sec in Mexico City and 27 m^3/sec in the State of Mexico (CNA, 1997a). About 7 per cent of this volume comes from 3,537 wells officially registered and operated by CNA and the governments of both Mexico City and the State of Mexico (Birkle et al., 1996). Legal wells are located in four different well fields in and around the ZMCM (National Research Council et al., 1995). It has not been possible, however, to calculate the exact volume of water abstracted from the aquifer owing to the existence of illegal wells, which could number 5,000–10,000 in the entire basin (Cruickshank, 1994; INEGI, 1996).

The management of drinking water supply, distribution, and wastewater collection within the ZMCM is shared by the governments of Mexico City and the State of Mexico. The political divisions of Mexican states are known as *municipios*, while the Distrito Federal is divided into 16 political *delegaciones*. These *municipios* and *delegaciones* are not autonomous. They do not establish their own guidelines for water planning and management purposes, nor do they decide how to maintain the water supply, distribution networks, or sewerage networks. All decisions are taken by the governments of both Mexico City and the State of Mexico.

At present, the CNA supplies about 24 m³/sec of water to the ZMCM. The CNA is also responsible for constructing and operating distribution systems to transfer water from other basins to the basin of the Valley of Mexico. It also operates some of the existing deep wells, while others belong to the State of Mexico and Mexico City governments (CNA, n.d.a).

Both Mexico City and the State of Mexico share water service areas, and each has five water service districts. Water enters the distribution system at specific points at one or more locations. The groundwater abstracted, the water withdrawn from wells, and the surface water from the few sources located within the basin go straight into the distribution system (National Research Council et al., 1995). This distribution system has become so big and complex that the water extracted from wells in one part of the ZMCM does not necessarily enter the system within the same service district.

At present, 97 per cent of the population in the D.F. and 90.5 per cent in the State of Mexico have access to water, either from a water connection directly to the house or from communal taps in the neighbourhood (National Research Council et al., 1995). However, most of the aquifers, springs, and rivers that supply water to the ZMCM are located in the west, north, and south of the area. Thus, the water supply is somewhat irregular and unreliable for the population living in the eastern part, who are the most affected by water shortages. More than 5 per cent of the people who live in the ZMCM still have to buy water from either public or private tankers. The cost of water per 200 litre container represents between 6 per cent and 25 per cent of their daily earnings (Restrepo, 1995). In 1994, poor people buying water from tankers were paying 500 times more than registered domestic consumers.

Main problems

The three existing sources from which water is abstracted for the ZMCM are the aquifer of the Valley of Mexico (71 per cent), the basins of the rivers Lerma–Balsas and Cutzamala (26.5 per cent), and the very few surface water bodies that still exist in the basin of the Valley of Mexico (2.5 per cent) (CNA, 1994, 1997b; UNAM, 1997). The rate of withdrawal from the aquifers is significantly higher than the recharge rate: 45 m³/sec is abstracted but the natural recharge rate is only 20 m³/sec, leaving an over-exploitation of 25 m³/sec (UNAM, 1997).

This over-exploitation has contributed to the depletion of the aquifer (the groundwater level is declining by about 1 m each year) and land subsidence at the rate of 10–40 cm/year in some parts of the city. It is estimated that the central area of ZMCM has subsided by 7.5 m during the past 100 years (World Resources Institute, 1996). The soil of Mexico City is basically clay and thus susceptible to dewatering and compaction. Accordingly, the higher the volume of water abstracted, the higher is the rate of land subsidence (CNA, 1997b). The sinking of the city has resulted in extensive damage to the city's infrastructures and water supply and sewer systems and degradation of the groundwater quality. It has also necessitated the construction of costly pumping plants to remove both wastewater and rainwater from the city (Departamento del Distrito Federal, 1991).

However, the problems related to water supply in the ZMCM extend beyond the sinking of the city. The entire hydraulic system, for example, has become not only very big and complex but also obsolete in many cases. Water distribution to the population varies in the different parts of the city, tariffs are still very highly subsidized, and the population wastes enormous amounts of water (people living in rich areas consume up to 600 litres, whereas people living in poor areas consume an average 20 litres).

The very high percentage of water that is lost through leaks in the distribution networks is due to the age of the pipes, the absence of adequate maintenance over the years, and continuing land subsidence and movement in the ZMCM. It is estimated that more than 30 per cent of water is lost in the pipes from leaks before reaching users (National Research Council et al., 1995; CNA, 1997b). The amount that is lost due to leaks would be enough to provide water to more than 4 million people (UNAM, 1997). Just in the ZMCM, the government repairs about 4,000 leaks in the distribution system every month (World Resources Institute, 1996). The efficiency of the water distribution systems all over Mexico leaves much to be desired; investment requirements are very high and so are the volumes of water that are wasted. In general, water losses due to leaks vary from a low of about 24 per cent to a high of 60 per cent in different cities in the country (table 5.2) (Arreguín-Cortés, 1994).

At present, one goal of the central and local governments is to promote water conservation within the population. One option would be to charge more realistic prices to domestic and industrial users. Services would not be as heavily subsidized as at present, and operational and maintenance costs would be met from the user fees. In

Table 5.2 **The results of some leakage studies in Mexico, 1991**

City	Volume supplied (litres/sec)	% of taps with leaks	% of supply lost from taps	% of supply lost in the supply system	Total losses (%)
Guaymas	488	30.0	23.4	1.8	26.23
Querétaro	1,783	14.0	13.5	2.8	29.96
Veracruz	2,869	17.0	24.2	0.1	24.34
Xalapa	1,215	9.0	34.4	8.9	43.32
Los Cabos	268	34.0	22.6	12.0	37.63
Oaxaca	721	24.0	59.2	1.1	60.34
Cancún	940	38.0	24.1	15.6	39.95
Chihuahua	3,489	5.0	15.8	25.7	41.50
Cd. Juárez	4,147	19.0	29.9	5.8	35.70
Average		21.1	27.4	8.2	37.66

Source: Arreguín-Cortés (1994).

1991, user charges represented only about 27 per cent of the operation and maintenance costs of the water supply systems (Departamento del Distrito Federal, 1991).

Until the middle of May 1997, the price charged to domestic users in Mexico City was US$0.2/m^3, when the cost of supplying water was about US$1.0/m^3. Since then, following a new pricing policy, tariffs have been reduced, depending on consumption, by 17–64 per cent (table 5.3). The Mexico City government claims that this "will contribute to better economy of the users and will strengthen public finances" (*Excelsior*, 24 May 1997). The fact remains that, as long as

Table 5.3 **Tariffs on average consumption in Mexico City, 1996 and 1997**

Consumption (m^3/2 months)	Consumption (litres)	Domestic tariffs		
		1996 (US$)	1997 (US$)	Difference (%)
30.1	500	10.25	3.75	−64
40.0	666	13.75	6.75	−51
50.0	833	17.12	9.75	−43
60.0	1,000	20.50	12.75	−38
60.1	1,001	24.62	12.75	−48
70.0	1,166	28.62	18.62	−35
80.0	1,333	32.75	24.62	−25
90.0	1,500	36.87	30.50	−17

Source: *Excelsior*, 24 May 1997.

the government keeps on subsidizing users, it will be very difficult to develop the necessary public awareness about water conservation and to reduce the amount of water that is wasted, and it will be almost impossible to recover the costs that are invested in operation and maintenance.

Water is charged per cubic metre as consumption level increases. There are numerous taps that are not registered, and thus consumption through them is neither recorded nor charged to the user. For example, in 1991, in Mexico City alone there were about 1.9 million taps, of which 1.3 million were registered; and, out of millions of users, only about 900,000 water meters were registered.

Macro-projects

In order to supply the necessary water to the most important cities in Mexico, the federal government is committed to building more water projects in the Valley of Mexico, Guadalajara, Monterrey, and Tijuana, starting in 1997, together with the local governments and the CNA. Even though the main objective is to solve the most acute problems related to drinking water supply and sanitation in these cities, the government will have to consider seriously in the foreseeable future the management of water demand, along with supply management. There is simply no other long-term alternative.

In the Valley of Mexico, the water projects mainly involve the enlargement of the Cutzamala distribution system; the construction of two aqueducts (Macrocircuit and "Aquaférico"); and covering 86 km of the currently uncovered main sewage line of the ZMCM. It was also planned to construct four wastewater treatment plants and to re-inject treated wastewater into the aquifer. However, in 1998, the newly elected government of Mexico City cancelled these plans. Several issues were considered. First, the investment costs for treating the wastewater and transferring it hundreds of kilometres from the source to the Mezquital Valley would be extremely high. Secondly, Mexico has not developed a cost-effective technology to treat wastewater and re-inject it into the aquifer. Thirdly, construction of just the treatment plants, without proper water resources management and planning, would not solve the acute sanitation problems of Mexico City and Mezquital Valley.

The total investment cost of these macro-projects would have been about US$1,800 million over a three-and-a-half-year period, including a US$500 million loan from the Interamerican Development

Bank (IDB), and financial assistance from Japan, the Mexican federal government, and the governments of the states of Mexico, Hidalgo, and Mexico City (*El Universal*, 21 May 1997).

The Cutzamala System

In 1976, the project known as the "Cutzamala System" (Sistema Cutzamala) was planned to supply water to the ZMCM from both the Cutzamala and the Lerma-Balsas river systems (in the State of Mexico) and to reduce the over-exploitation of the aquifer of the Valley of Mexico (CNA, 1997b). The Cutzamala System is the second source of water for the ZMCM. It supplies water to the north of Mexico City and to the State of Mexico. The water has to be transferred from 60–154 km away and pumped to a height of more than 1,000 m, thus requiring 102 pumping stations, which makes this operation extremely energy intensive and expensive (SEDUE, 1990; CNA, 1997b).

Because of the magnitude of the project, its construction was initially planned in three stages. The first came into operation in 1982 (4 m^3/sec), the second in 1985 (6 m^3/sec), and the third in 1993 (9 m^3/sec) (CNA, n.d.b). In 1997, a fourth stage (the Temascaltepec project) was expected to be initiated. However, the government has not been able to start construction on the project owing to severe social problems (CNA, 1997b). People living in the areas that will be affected by the construction of the fourth stage have opposed the project, because they think it will supply water to the people living in Mexico City, and there is no reason why they should suffer simply because water is needed in another part of the country. To a significant extent this is due to the lack of an adequate strategy on information and communication from the governments to society and to the absence of proper public participation – the project will in fact benefit people living both in Mexico City and in the State of Mexico (CNA, personal communication). Government institutions have generally not considered the potential social conflicts resulting from the transfer of water resources from one basin to another; nor have they properly analysed the nature of the beneficiaries and the people who may have to pay the cost. In fact, not even the Environmental Impact Statement (EIS) for the fourth stage of the Cutzamala System (CNA, 1997b) mentions the social implications. Like most EISs in Mexico, it considers almost exclusively technical factors; social issues are conspicuous by their absence.

The Cutzamala System utilizes seven reservoirs and comprises a pipeline for drinking water, a regulatory reservoir, and a 127 km aqueduct, which includes 21 km of tunnels, a 7.5 km open canal, one water treatment plant (24 m^3/sec capacity) (CNA, n.d.c), and six pumping stations to raise the water 1,300 m, which requires total energy of 1,650 kWh/year (CNA, 1997b). The water is first treated at source in the Los Berros treatment plant (pre-chlorination, alum coagulation/flocculation, gravity sedimentation, and rapid sand filtration) and then it enters the Cutzamala System (National Research Council, 1995).

Initially, what was later converted into the Cutzamala System was a hydropower project. Cutzamala took advantage of the infrastructure that already existed for the hydropower, but the planned water use was changed. Currently, only 3 m^3/sec is used to generate hydropower during peak hours and to satisfy local energy requirements in the agricultural and industrial energy sectors (CNA, 1997b). The programme on drinking water, drainage, and sanitation of the ZMCM now expects to increase the water supply from the Cutzamala System to the Valley of Mexico from 0.6 km^3/year (19 m^3/sec) to 0.76 km^3/year (24 m^3/sec), and to treat 1.3 km^3/year (42 m^3/sec) of wastewater (CNA, 1994, 1997b).

According to the EIS carried out for the fourth stage, the total investment cost of the first three stages was US$965 million (1996 estimates). If the estimated cost of the facilities from the cancelled hydroelectric system is added, the total investment cost becomes US$1,300 million. The reservoirs of the earlier hydroelectric plants represent a volume of 840 million m^3 (CNA, 1997b).

The total area affected by the construction of the Cutzamala System during the first three stages is approximately 710 ha, with a land value of US$3.55 million (CNA, 1994, 1997b). One of the main adverse socio-economic impacts of the Cutzamala System has been the resettlement of communities, who, as of February 1999, had not received the expected compensation.

In addition to the construction of the Cutzamala System, about 190 so-called social projects have been built to benefit some of the people living in the most affected municipalities (CNA, 1994, 1997b). These projects were built jointly by CNA and the communities, and consist mainly of the construction, enlargement, and rehabilitation of both water supply and sanitation systems, as well as the construction and rehabilitation of houses, schools, and farms. Equally important is the construction and rehabilitation of roads by the CNA, both for the

Cutzamala System and for social benefit. The cost of these social projects was estimated in 1996 to be equivalent to 5 per cent of the direct investment in the Cutzamala System, which would represent an additional US$45 million (1996 estimates) (CNA, 1997b).

It is worth noting that the total cost of the Cutzamala System at US$1,300 million (mainly construction and equipment costs) was higher than national investment in the entire public sector in Mexico in 1996, including education (US$700 million), health and social security (US$400 million), agriculture, livestock, and rural development (US$105 million), tourism (US$50 million), and the marine sector (US$60 million). Up to 1994, the Cutzamala System alone represented three times the annual infrastructural expenditure of the Ministry of Environment, Natural Resources and Fisheries for 1996, which was more than US$470 million (CNA, 1997b).

The energy requirement to run the Cutzamala System is about 1,787 million kWh/year, which represents an approximate cost of US$62.54 million. The total investment cost would increase significantly if the investment costs in personnel (US$1.5 million/year) as well as the water treatment process costs were added (CNA, 1997b). The energy consumed by the system, plus the energy that could have been produced if the hydroelectric system had continued to be used as originally planned, could have supplied power to a population of about 2.59 million people.

If only the operational costs for running the Cutzamala System are considered (about US$128.5 million/year), supplying 600 million m^3 of water (19 m^3/sec) would mean an average cost per m^3 of US$0.14 and energy consumption of 6.05 kWh/m^3. The latter figure is more than seven times the consumption of power of locations near the ZMCM. The price of water of about US$0.2/$m^3$ is not enough to cover either the operational costs of the Cutzamala System or the purification or distribution costs of water to the ZMCM. According to the EIS (CNA, 1997b) for the fourth stage of the Cutzamala System, the minimum price of water to cover expenses should be over US$0.3/$m^3$. It would be even higher if the cost of treating and distributing the water is included.

Once the fourth stage of the Cutzamala System is operating, the water supplied will increase from 19 m^3/sec to 24 m^3/sec. This last stage includes the construction of a reservoir with a capacity of 65 million m^3 to regulate an approximate flow of 5,000 litres/sec, a 15 m^3/sec pumping station, 18 km of canals, and 12 km of tunnels (CNA, 1997b).

Some studies have indicated that, if the leaks in the distribution system from Cutzamala to the ZMCM were repaired, there would be no need to construct the fourth stage of the project. This means that the additional water supply of 5 m^3/sec that is being planned at very high investment, social, and environmental costs would not have been necessary had better planning and management practices been implemented. The government has so far made no public statement on this issue.

The Cutzamala Macrocircuit and the Cutzamala "Aquaférico"

The federal government, together with the government of the State of Mexico and the CNA, is constructing two distribution lines in order to facilitate a better distribution system for the water coming from the Cutzamala System.

The Federal District is constructing an aqueduct, known as "Aquaférico," that will distribute water from the Cutzamala System, which comes from the west, to the southern and eastern parts of the ZMCM (National Research Council et al., 1995; CNA, 1997d).

In the State of Mexico, the water distribution system is an aqueduct known as the "Cutzamala Macrocircuit." It is expected to be completed by the year 2000 and will be built around the perimeter of most of Mexico City to the north, carrying water to the northern, southern, and eastern parts of the city (CNA, n.d.d,e,f). The first stage of the Macrocircuit was inaugurated in October 1994. Both the first and the second stages of the Macrocircuit are now in operation, providing a continuous water supply of 4 m^3/sec and benefiting 1,382,400 people by 250 litres/day/person. The third and fourth stages of the Macrocircuit will increase the drinking water supply by an additional 7 m^3/sec (making a total volume of 11 m^3/sec), benefiting 4,752,000 inhabitants living in the eastern and northern areas of the State of Mexico by about 200 litres/day/person (CNA, n.d.d,e,f; CNA, 1994, 1997c). The Macrocircuit includes the construction of two pipelines of 168.28 km which will connect to pipelines already built, making a total of 226.48 km. The two pipelines will need a surface area of 336.56 ha, plus 71 ha for the storage tanks (CNA, 1997c).

Total investment in the Macrocircuit between 1987 and 1997 was US$78 million, and the estimated cost for stages three and four (1997–2000) is expected to be about US$190 million, making a total investment of US$268 million. This amount represents almost 50 per cent of the total budget of the public sector at the national level for

1995 (US$563 million) in the areas of urban development, ecology, and drinking water (CNA, 1997c).

Sewage

Between the beginning of the twentieth century and 1936, parts of Mexico City were sinking by about 5 cm/year. However, higher water demand resulted in the construction and operation of deeper wells between 1938 and 1948, which increased the land subsidence rate first to 10 cm/year, and later to 30 to 40 cm/year. The functioning of the sewerage system, which until then worked on a gravity basis, was severely affected by this settlement. The network also sank with the city and its level in the different parts of the city became uneven. Consequently, it became necessary to pump wastewater up from the small sewage pipes to the level of the main wastewater collector of the city, increasing both maintenance and operation costs.

However, the increasing population in the ZMCM made the sewage collection and treatment capacity insufficient. It was then decided to build another main collector for wastewater for both Mexico City and the State of Mexico as a combined sewage and rainwater network ("Drenaje profundo"). This system is constructed up to 300 m below the ground level of the city. It is thus not affected by the sinking of the city (Departamento del Distrito Federal, 1990)

This main collector carries both rainwater (an annual average of 14 m^3/sec) and wastewater (48 m^3/sec) through primary and secondary networks. The secondary network is used to transport municipal and industrial wastewater and rainwater in pipes of up to about 6 m in diameter. The primary network is connected to the secondary network and stores, transports, and disposes of the wastewater into the Gulf of Mexico through four artificial channels located at the northern end of the basin (UNAM, 1997; National Research Council et al., 1995). The networks have 66 pumping stations, regulatory tanks for flow control, storm tanks, 111 km of open canals, piped rivers, dams, and lagoons, and 118 km of underground collectors and tunnels. According to 1995 figures, the total volume of wastewater discharged into the ZMCM sewerage system was 2,349,116 km^3 (Departamento del Distrito Federal, 1990; National Research Council et al., 1995; INEGI, 1996).

Being located within a naturally closed hydrological basin, the city is especially vulnerable to flooding. Throughout history, artificial channels have been built to carry wastewater mixed with rainwater out of the city. The rainy season in the ZMCM is characterized by storms of short duration and high intensity, which can produce up to 70 mm of rainfall, representing 10 per cent of the total annual precipitation. Thus, the main collector was designed to carry about 200 m^3/sec of water over a 45-hour period, even though it has carried up to 340 m^3/sec (National Research Council et al., 1995). Such sudden fluctuations in the amounts of water that have to be drained create severe problems for the design and operation of the infrastructure.

The sinking of the city has also affected the sewerage system owing to pipe fractures and loss of the hydraulic gradient, which has significantly reduced the efficiency of the whole urban sewerage system as well as contributing to groundwater contamination. New investments include covering 86 km of the currently unlined and open main collector, which would prevent the dumping of garbage and also eliminate environmental and health risks.

Constraints on water resources management

The management of water resources in the ZMCM is very complex. There seems to be an uncontrolled race between the water and wastewater needs of an increasing population and the budget, technology, and management expertise to construct, operate, and maintain all the necessary systems efficiently.

The problems of water quantity and quality in the ZMCM are linked to regional economic development policies and continual increases in population. Government policies in recent years have attempted to promote the development of other urban centres to alleviate poverty and to provide better standards of living as well as quality of life. The policy appears to be working, since the migration rates to the ZMCM as a whole have declined in recent years. However, even though the population growth rate in Mexico City during the early part of the twenty-first century is expected to decline to 2.1 per cent per year, the growth rates in the municipalities of the State of Mexico around Mexico City are expected to increase even further (Cruickshank, 1994), which really will not solve the overall problem.

Unless current trends change, the future scenario will include very high investment costs to transport more and more water from increasingly distant and expensive sources, greater land subsidence owing to increasing groundwater withdrawal, a reduction in the quality of the water extracted from the aquifer, higher subsidies, and higher investments to cover operation and maintenance costs, etc. The result can only be a "lose–lose" situation for everyone concerned.

Another constraint stems from the fact that the demand for living space from the continually increasing population of the ZMCM has forced land-use changes. Concrete and asphalt now cover areas that are essential for the recharge of the aquifers. For example, the southern area of the city is a good area for aquifer recharge, because the soil is broken basalt. However, it is now heavily urbanized, and this is also one of the main sources of groundwater contamination because of the absence of a sewerage system (which cannot be economically constructed owing to the presence of volcanic rock) (CNA, 1997b). Houses are thus built only with septic tanks. The change of land use also contributes to higher volumes of rainwater, which enter the sewerage system, requiring higher capacities.

The risk of contamination of the aquifer is increased owing to the disposal of untreated industrial wastewater directly into the sewerage system, inadequate waste treatment facilities, leaks from sewage pipes, and waste illegally dumped in landfills and in unlined sewage canals (World Resources Institute, 1996). The 1990 census indicates that some 82 per cent of the houses in the ZMCM are connected to the sewerage system, 6 per cent use septic tanks, and 12 per cent discharge their solid and liquid wastes directly onto the land or into the water (CNA, 1997c). In the State of Mexico, poor-quality drinking water has been detected in domestic taps as a result of the infiltration of unclean water into a leaky distribution system and the precipitation of salts (mainly calcium, magnesium, iron, and manganese) (National Research Council et al., 1995). Water contamination has very serious public health impacts. The gastro-enteric diseases that result from the consumption of polluted water are the second major cause of child mortality (278 per 100,000) in Mexico (UNAM, 1997), the third leading cause in the State of Mexico (450 per 100,000), and the fourth in Mexico City (157 per 100,000) (National Research Council et al., 1995).

In terms of water reuse, most of this occurs informally through the use of raw wastewater for irrigation. The wastewater from the city ends up in Endhó Dam, in the State of Hidalgo, 109 km north of Mexico City, where it is used for irrigation purposes. The agricultural potential of this area, the Mezquital Valley, used to be very poor because of the semi-arid climate. However, the use since 1912 of wastewater irrigation for agricultural production in this valley has significantly improved production yields. This area is now known as the "bread basket" of the country, with more than 5,000 ha irrigated

(Gutiérrez-Ruiz et al., 1995). Currently there is a decree that limits the wastewater that can be delivered to the Mezquital Valley to a maximum of 400 million m^3/year. In spite of this, different sources affirm that the valley receives about 1,700 million m^3/year and the farmers have been promised that they will receive all the wastewater originating in the Valley of Mexico (Gutiérrez-Ruiz et al., 1995).

Because of the salinity of the wastewater, its productivity is very low. This is compensated for by over-irrigation. The main crops grown in the Mezquital Valley are alfalfa and corn, which account for 60–80 per cent of the total cultivated area. The cultivation of vegetables that are consumed raw is forbidden by law, but the law is not always followed (Gutiérrez-Ruiz et al., 1995). Another concern is the infiltration of contaminated water into the unlined channels, which is why the open channels are to be replaced by pipes.

Irrigation with wastewater has been beneficial because it has provided additional nutrients to the soil and crops and because it has also supplied water to a semi-arid region. However, it poses a very high risk to the health not just of the people who live and work in the irrigation districts but also of consumers (National Research Council et al., 1995; SEDUE, 1990).

The proposal to re-inject treated wastewater into the aquifer raised some environmental and health concerns. At present, the Endhó Reservoir is completely covered with water hyacinths, with consequent environmental and health risks to the population living nearby. This is in addition to the overall health issues raised by irrigation with raw wastewater. The farmers in Mezquital Valley are already worried about the potential economic side-effects they may suffer if water is re-injected into the aquifer. It would mean that they might not receive the same quantity of water as they are receiving at present. The quality of the water would also change, because the nutrient content of properly treated wastewater would be lower. This means the farmers would have to increase their use of fertilizers, thus increasing their production costs.

An aquifer recharge programme to reduce flooding was started in 1943 through runoff retention, surface spreading, channel modification, and infiltration wells. An artificial recharge programme using injection wells was initiated in Mexico City in 1953. However, the quality of the water was not monitored, and wells had subsequently to be closed as a result of contamination problems (National Research Council et al., 1995). The future scenario is thus not very optimistic. Before implementing any programme to inject treated wastewater into the aquifer of the Valley of Mexico, many precautions would be necessary. Equally, current users of wastewater in the Mezquital Valley need to be consulted properly. Otherwise it could contribute to social conflicts.

Other activities that reuse treated wastewater include the watering of green areas for recreational activities, the irrigation of farmland, and filling up lakes. This reuse amounts to about 4 m^3/sec of treated wastewater from Mexico City. In the State of Mexico, most of the treated wastewater is reused in industry (UNAM, 1997). The government has given concessions for operating treatment plants to the private sector, which is also considering potential users of the treated wastewater (Departamento del Distrito Federal, 1991). The government has also been promoting the use of more efficient water closets, which could save more than 70 million litres/day. In addition, the government of Mexico City has appropriated land to create more green areas, thus contributing to aquifer recharge (Departamento del Distrito Federal, 1991; UNAM, 1997).

Conclusions

Clearly the present approach to the management of the water supply and wastewater in the Valley of Mexico is neither efficient nor sustainable. In order to meet the needs of the population in the ZMCM in terms of water quantity and quality, and simultaneously to maintain a proper balance between the people, natural resources, the environment, and health, it is necessary to develop and implement an integrated management plan that explicitly considers the interests of the different sectors as well as appropriate economic, social, technical, political, environmental, and institutional factors. The need for public consultation and involvement in preparing and implementing such a plan should not be underestimated.

The CNA has developed a plan that is currently under review. It is known as the "Project on Sanitation for the Valley of Mexico," which is expected to comply with environmental policies by the year 2000. The complexity of the problem in the ZMCM will eventually force the authorities concerned to search beyond purely engineering solutions and the construction of infrastructure without simultaneously considering appropriate demand management strategies.

A Basin Council for the Valley of Mexico already exists, and it will have to work with the representatives of the federal government, governments from the states of Mexico and Hidalgo, the CNA, and other institutions associated with local decision-making in the area of water planning and management. The institutions concerned will have to realize that environmental and social issues are important considerations when developing and implementing long-term strategies for managing water supply and wastewater disposal. Realistic policies and programmes on water use, reuse, and conservation will have to be implemented within the next decade. Massive efforts will be needed to enhance public awareness and understanding of the seriousness of the water problem, and the role the public must play in its resolution. Issues such as cost recovery and appropriate levels of water pricing can no longer be ignored. For example, at the present average price for water for domestic use in the State of Mexico of about US$0.20/m^3, it would take over 25 years to recover the investment in the Macrocircuit (CNA, 1997c).

The future of the ZMCM does not look very promising. Clearly some hard decisions will have to be taken in the near future. It is no longer a feasible long-term option continually to increase investment in water supply and wastewater treatment in the ZMCM at the expense of the people living in other parts of the country, or at the expense of other types of social investment. The problem is complex but, given the political will, it can be resolved. It will not be an easy task, however.

References

Arreguín-Cortés, F. 1994. "Uso Eficiente del Agua en Ciudades e Industrias," in *Uso Eficiente del Agua*, edited by H. Garduño & F. Arreguín-Cortés, CNA, IMTA, UNESCO-ORCYT, IWRA, Mexico, pp. 63–91.

Birkle, P., Torres-Rodríguez, V., and González-Partida, E. 1996. "Balance de Agua de la Cuenca del Valle de México y su Amplicación para el Consumo en el Futuro," in *III Congreso Latinamericano de Hidrología Subterránea, Proceedings*, Asociación Latinoamericana de Hidrología Subterránea para el Desarrollo, San Luis Potosí, México, pp. 113–124.

Casasús, C. 1994. "Una Nueva Estrategia para la Ciudad de México," *Agua*, Comisión de Aguas del Distrito Federal, December, pp. 9–18.

CNA (Comisión Nacional del Agua). n.d.a. *Planta Potabilizadora Madín*, Comición Nacional del Agua, Gerencia de Aguas del Valle de México, Unidad de Información y Participación Ciudadana, México.

——— n.d.b. *Subsistema Chilesdo, Tercera Etapa Sistema Cutzamala*, Comisión Nacional del Agua, Gerencia de Aguas del Valle de México, Unidad de Información y Participación Ciudadana, México.

—— n.d.c. *Planta Potabilizadora Los Berros, Sistema Cutzamala*, Comisión Nacional del Agua, Gerencia de Aguas del Valle de México, Unidad de Información y Participación Ciudadana, México.

—— n.d.d. *Sistema Cutzamala, Ramal Norte Macrocircuito, I Etapa*, Comisión Nacional del Agua, Gerencia de Aguas del Valle de México, Unidad de Información y Participación Ciudadana, México.

—— n.d.e. *Sistema Cutzamala, Ramal Norte Macrocircuito, II Etapa*, Comisión Nacional del Agua, Gerencia de Aguas del Valle de México, Unidad de Información y Participación Ciudadana, Mexico.

—— n.d.f. *Sistema Cutzamala, Ramal Norte Macrocircuito, III Etapa*, Comisión Nacional del Agua, Gerencia de Aguas del Valle de México, Unidad de Información y Participación Ciudadana, México.

—— 1994. *Informe 1989-1994*, Internal Report, Comisión Nacional del Agua, Secretaría de Agricultura y Recursos Hidráulicos, México.

—— 1997a. *Situación del Subsector Agua Potable, Alcantarillado y Saneamiento a diciembre de 1995*, Comisión Nacional del Agua, México.

—— 1997b. *Diagnóstico Ambiental de las Etapas I, II y III del Sistema Cutzamala*, Comisión Nacional del Agua, México.

—— 1997c. *Manifestación de Impacto Ambiental Modalidad Específica del Proyecto Macrocircuito Cutzamala*, Comisión Nacional del Agua, México.

—— 1997d. *Estrategias del Sector Hidráulico*, Comisión Nacional del Agua, México.

CONAPO. 1997. *Población de la Zona Metropolitana de la Ciudad de México, 1970-2050. Estimaciones y Proyecciones del Consejo Nacional de Población*, México.

Cruickshank, G. 1994. *Proyecto Lago de Texcoco, Rescate Hidrológico*, Comisión Nacional del Agua, México.

Departamento del Distrito Federal. 1990. *El Sistema de Drenaje Profundo de la Ciudad de México*, Secretaría General de Obras, Dirección General de Construcción y Operación Hidráulica, 2nd edn., México.

—— 1991. "Estrategia para la Ciudad de México," *Agua 2000*, Mexico.

Gutiérrez-Ruiz, M. E., Siebe, C., and Sommer, I. 1995. "Effects of Land Application of Wastewater from Mexico City on Soil Fertility and Heavy Metal Accumulation: A Bibliographical Review," *Environmental Review*, Vol. 3, pp. 318-330.

INEGI. 1996. *Anuario Estadístico del Distrito Federal*, Instituto Nacional de Estadística, Geografía e Informática, México.

Naranjo, F. and Biswas, A. K. 1997. "Water, Wastewater and Environmental Security: A Case Study of Mexico City and Mezquital Valley," *Water International*, Vol. 22, No. 3, September.

National Research Council, Academia de la Investigación Científica, A.C. and Academia Nacional de Ingeniería, A.C. (eds.). 1995. *Mexico City's Water Supply: Improving the Outlook for Sustainability*, National Academy Press, Washington, D.C.

Restrepo, I. (ed.). 1995. *Agua, Salud y Derechos Humanos*, Comisión Nacional de Derechos Humanos, Mexico.

SEDUE. 1990. *Control de la Contaminación del Agua en México*, Subsecretaría de Ecología, Dirección General de Prevención y Control de la Contaminación Ambiental, México.

SEMARNAP/CNA. 1996. *Programa Hidráulico 1995–2000*, Poder Ejecutivo Federal, Estados Unidos Mexicanos, México.

UNAM (ed.). 1997. *Environmental Issues: The Mexico City Metropolitan Area*, Programa Universitario del Medio Ambiente, Departamento del Distrito Federal, Gobierno del Estado de México, Secretaría de Medio Ambiente, Recursos Naturales y Pesca, México.

World Resources Institute. 1996. "Water: The Challenge for Mexico City," in *World Resources. A Guide to the Global Environment 1996–1997*, Oxford University Press for the World Bank, New York.

6

Wastewater management and reuse in mega-cities

Takashi Asano

Introduction

Electricity, gas, water, sewer, and garbage collection services are traditionally provided by municipalities. The extent of direct municipal involvement, however, often differs from one utility to another and from one community to another, although relatively few cities generate electricity – most municipalities that provide electrical services purchase electricity and resell it to their customers. On the other hand, many municipalities assume a more comprehensive and direct role in the treatment and distribution of potable water and the collection and treatment of wastewater.

In reviewing the problems of wastewater management in mega-cities, the subject of this chapter, one notes that the socio-economic conditions confronting developing countries in the twenty-first century are overwhelming. The almost exponential growth in the number of large cities with more than 5 million people has been reported: in 1950 there were 6 such cities, in 1980 there were 26 mega-cities; in the year 2000, however, there may be 60; and by 2025 they are expected to grow to 90. As a consequence, water problems will create escalating conflicts between different parts of the urbanized area and its surroundings (Lindh, 1992) as well as mounting problems related to wastewater treatment and disposal.

A water supply and a sanitation system are the primary infrastructure needs of any community, particularly in mega-cities, playing a key role in providing amenities to the people, protecting the environment, and eliminating water-borne diseases. Thus, a well-functioning urban drainage (sewerage) and wastewater treatment system is the most effective solution to sewage and urban runoff problems as well as to maintaining and enhancing healthy living conditions in mega-cities.

Ultimately, after appropriate treatment, wastewater collected from cities must be returned to the land or water. The complex question of which contaminants in urban wastewater should be removed to protect the environment, to what extent, and where they should be placed must be answered in light of an analysis of local conditions, environmental risks, scientific knowledge, engineering judgement, and economic feasibility. Providing these infrastructures is, however, costly. For example, the total, 20-year capital cost to upgrade US municipal sewerage systems is estimated to be US$110 billion (for the design year of 2010 in 1990 dollars) according to the 1990 Needs Survey Report to the Congress. The cost for constructing conventional secondary (US$37.3 billion) and advanced (US$11.79 billion) wastewater treatment systems totals US$49.0 billion (National Research Council, 1993).

Given the escalating conflicts in water resources development and recognizing that complete sewerage construction will not be possible in the foreseeable future, it is clear that wastewater reclamation and reuse will play a central role in future long-term strategies for the management of water resources and wastewater. Wastewater reclamation and reuse can serve several objectives. The most prominent can be summarized as providing: (1) a water supply to displace the need for other sources of water, (2) a cost-effective means for the environmentally sound treatment and disposal of urban wastewater, and (3) an incidental secondary benefit from the disposal of wastewater, primarily crop production by irrigation.

Reclaimed wastewater is, after all, a water resource developed right on the doorstep of the urban environment where water resources are needed the most and priced the highest. Furthermore, reclaimed wastewater provides a reliable source of water even in drought years because the generation of urban wastewater is little affected by drought.

The role of wastewater reuse in mega-cities

Reflecting the experience and primary motivations in wastewater reclamation and reuse in the United States, the emphasis of this chapter will be on the first and second objectives listed above, that is, wastewater reuse planned as a water supply to offset alternative water development and as a cost-effective means of municipal wastewater pollution control.

Today, technically proven wastewater treatment or water purification processes exist to provide water of almost any quality desired. However, the effective integration of water and reclaimed wastewater still requires close examination of public health issues, infrastructure and facilities planning, wastewater treatment plant siting, treatment process reliability, economic and financial analyses, and water utility management. Whether wastewater reuse can be implemented will depend upon careful economic considerations, potential uses for the reclaimed water, stringency of waste discharge requirements, public health considerations, and public policy emphasizing water conservation rather than the construction of new water resources facilities.

The inclusion of planned wastewater reclamation, recycling, and reuse in water resource management systems reflects the increasing scarcity of water sources to meet societal demands, technological advancement, public acceptance, and improved understanding of public health risks. As the link between wastewater, reclaimed water, and water reuse has become better understood, increasingly smaller recycle loops can be developed. Traditionally, the hydrological cycle has been used to represent the continuous transport of water in the environment. The water cycle consists of fresh and saline surface water resources, sub-surface groundwater, water associated with various land-use functions, and atmospheric water vapour. Many sub-cycles to the large-scale hydrological cycle exist, including the engineered transport of water such as aqueducts.

In urban, industrial, and agricultural areas, wastewater reclamation, recycling, and reuse have become significant components of the hydrological cycle. A conceptual overview of the cycling of water from surface and groundwater resources to water treatment facilities, to irrigation, municipal, and industrial applications, and to wastewater reclamation and reuse facilities is shown in figure 6.1.

Water reuse may involve a completely controlled "pipe-to-pipe" system with an intermittent storage step, or it may include blending of reclaimed water with natural water either directly in an engineered

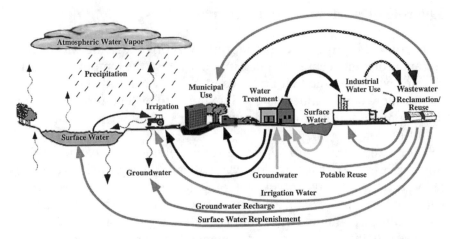

Fig. 6.1 **The role of engineered treatment, reclamation, and reuse facilities in the cycling of water through the hydrological cycle (Source: after Asano and Levine, 1995)**

system or indirectly through surface water supplies or groundwater recharge. The major pathways of water reuse are depicted in figure 6.1 and include groundwater recharge, irrigation, industrial use, and surface water replenishment. Surface water replenishment and groundwater recharge also occur through natural drainage and through infiltration of irrigation and stormwater runoff. The potential use of reclaimed wastewater for potable water treatment is also shown. The rate and quantity of water transferred via each pathway depend on watershed characteristics, climatic and geohydrological factors, the degree of water utilization for various purposes, and the degree of direct or indirect water reuse (Asano and Levine, 1995; Asano, 1998).

The water used or reused for agricultural and landscape irrigation includes agricultural, residential, commercial, and municipal applications. Industrial reuse is a general category encompassing water use for a diversity of industries that include power plants, food processing, and other industries with high rates of water utilization. In some cases, closed-loop recycling systems have been developed that treat water from a single process stream and recycle the water back to the same process with some additional make-up water. In other cases, reclaimed municipal wastewater is used for industrial purposes such as in cooling towers. Closed-loop systems are also under evaluation for reclamation and reuse of water during long-duration space missions by the National Aeronautics and Space Administration (NASA).

Overview of wastewater reclamation technologies

The effective treatment of wastewater to meet water quality objectives for water reuse applications and to protect public health is a critical element of water reuse systems. Conventional municipal wastewater treatment consists of a combination of physical, chemical, and biological processes and operations to remove solids, organic matter, pathogens, toxic metals, and sometimes nutrients from wastewater. General terms used to describe different degrees of treatment, in order of increasing treatment level, are preliminary, primary, secondary, tertiary, and advanced treatment. A disinfection step for the control of pathogenic organisms is often the final treatment step prior to distribution or storage of reclaimed wastewater. Wastewater reclamation technologies are largely derived from applying technologies used for conventional wastewater treatment and drinking water treatment. The goal in designing a wastewater reclamation and reuse system is to develop an integrated cost-effective treatment scheme that is capable of reliably meeting water quality objectives.

To meet that goal, water quality characterization is necessary to evaluate the biological and chemical safety of using reclaimed water for various applications and the effectiveness and reliability of individual treatment technologies. The water quality parameters that are used to evaluate reclaimed wastewater are based on current practice in water and wastewater treatment. A summary of relevant water quality monitoring parameters is given in table 6.1. Municipal wastewater treatment systems are typically designed to meet water quality objectives based on biochemical oxygen demand (BOD_5), total suspended solids (TSS), total or faecal coliforms, nutrient levels (nitrogen and phosphorus), and chlorine residual. Potable water quality monitoring parameters include micro-organisms such as coliforms, turbidity, dissolved minerals, disinfection by-products, and specific inorganic and organic contaminants. Recently, there has been increased emphasis on developing monitoring tools for the detection of microbial pathogens including *Giardia lambia, Cryptosporidium parvum*, and enteric viruses in potable water supplies. Particle-size analysis of wastewater contaminants has also been proposed as a water quality monitoring tool.

Where wastewater is not treated to the secondary level (secondary

Table 6.1 **Summary of the major parameters used to characterize reclaimed wastewater quality**

Parameter	Significance in wastewater reclamation	Approximate range in treated wastewater	Treatment goal in reclaimed wastewater[a]
Organic indicators			
BOD_5	Organic substrate for microbial or algal growth	10–30 mg/litre	<1–10 mg/litre
TOC	Measure of organic carbon	1–20 mg/litre	<1–10 mg/litre
Measurement of particulate matter			
Total suspended solids (TSS)	Measure of particles in wastewater can be related to microbial contamination and turbidity; can interfere with disinfection effectiveness	<1–30 mg/litre	<1–10 mg/litre
Turbidity	Measure of particles in wastewater; can be correlated to TSS	1–30 NTU	0.1–10 NTU
Pathogenic organisms	Measure of microbial health risks due to enteric viruses, pathogenic bacteria, and protozoa	Coliform organisms: <1 to 10^4/100 ml Other pathogens: controlled by treatment technology	<1–2,000/ml
Nutrients			
Nitrogen	Nutrient source for irrigation; can also contribute to microbial growth	10–30 mg/litre	<1–30 mg/litre
Phosphorus	Nutrient source for irrigation; can also contribute to microbial growth	0.1–30 mg/litre	<1–20 mg/litre

a. Treatment goal depends on specific wastewater reuse application.

treatment) and not effectively disinfected, the major health risk is exposure to biological agents, including bacterial pathogens, helminths, and protozoa. From a process control and public health perspective in wastewater reclamation and reuse, enteric viruses are the most critical group of pathogenic organisms in the developed world owing to the possibility of infection from exposure to low doses and the lack of routine, cost-effective methods for the detection and quantification of viruses. In addition, treatment systems that can remove enteric viruses effectively will be effective for the control of the other pathogenic organisms listed above.

In the mid-1970s, concerns about pathogen removal, particularly enteric viruses, from treated effluent discharged to ephemeral streams in southern California prompted the adoption of conventional water treatment processes as a tertiary wastewater treatment system in both discharge to urban waterways and high-exposure water reuse applications. To achieve essentially virus-free effluent, the "full treatment" for wastewater reclamation stipulated by the California Department of Health Services consisted of coagulation, flocculation, sedimentation, filtration, and disinfection (with chemical doses of approximately 150 mg/litre alum, 0.2 mg/litre polymer, and 10 mg/litre chlorine). This process train is often referred to as the "Title 22 process" owing to its inclusion as the reference treatment train in Title 22 of the California Code of Regulations, *Wastewater Reclamation Criteria* (State of California, 1978).

Because the "Title 22 process" is costly (as shown in table 6.5 below), the "direct filtration" tertiary treatment system is now in common use in California. This consists of granular-medium filtration (with or without the addition of a small quantity of alum in the range of 2–5 mg/litre) and chlorine disinfection (with approximately 10 mg/litre chlorine dose and 1.5 hour contact time). Considerable efforts have been directed to the development of tertiary treatment systems alternative to "Title 22 process," notably in the Pomona Virus Study (Dryden et al., 1979) and the Monterey Wastewater Reclamation Study for Agriculture (Sheikh et al., 1990). The "direct filtration" process has been found to be particularly suited as a tertiary treatment system when good-quality secondary effluent is available. An increasing number of water reuse facilities using a direct filtration tertiary system have been constructed for water reuse applications in the urban environment such as in parks, school yards, and golf courses. More than 40 such tertiary filters are in use in California, specifically installed for wastewater reclamation purposes (State of California, 1990).

As noted above, the degree of treatment required in individual water treatment and wastewater reclamation facilities varies according to the specific reuse application and associated water quality requirements. The simplest treatment systems involve solid/liquid separation processes and disinfection, whereas more complex treatment systems involve combinations of physical, chemical, and biological processes employing multiple barrier treatment approaches for contaminant removal. An overview of the major technologies that are appropriate for wastewater reclamation and reuse systems is given in table 6.2.

Wastewater reuse applications

To provide a framework for evaluating wastewater reuse, it is important to correlate major water use patterns with potential water reuse applications. On the basis of water quantity, irrigation use (consisting of both agricultural and landscape applications) is projected to account for 54 per cent of total freshwater withdrawals in the United States by the year 2000. The second major user of reclaimed water is industry, primarily for cooling and processing needs. However, industrial uses vary greatly, and additional wastewater treatment beyond secondary treatment is usually required. Thus, the effective integration of wastewater reuse into water resource management is based on the quantity of water required for a specific application and the associated water quality requirements.

Significant regional and seasonal variations in water use patterns exist. For example, in urban areas, industrial, commercial, and non-potable urban water requirements account for the major water demand. For agricultural applications in arid and semi-arid regions, irrigation is the dominant component of water demand. Water requirements for irrigation applications tend to vary seasonally, whereas industrial water needs are more consistent. The feasibility of water reuse for a given watershed is limited by the degree to which reclaimed wastewater could augment existing water supplies through substitution of water in commercial, industrial, and agricultural applications.

An overview of the major categories of wastewater reuse is given in table 6.3. These categories are arranged according to current and projected volumes of reclaimed wastewater. Treatment goals are based on effluent quality and the application of specific technologies. For most applications, effective secondary treatment is a prerequisite to the production of high-quality reclaimed water. The primary incentives for the implementation of water reuse are related to the need to augment water supplies or control water pollution in receiving

Table 6.2 **Overview of representative unit processes and operations used in wastewater reclamation**

Process	Description	Application
Solid/liquid separation		
Sedimentation	Gravity sedimentation of particulate matter, chemical floc, and precipitates from suspension by gravity settling	Removal of particles from wastewater that are larger than about 30 μm. Typically used as primary treatment and downstream of secondary biological processes
Filtration	Particle removal by passing water through sand or other porous medium	Removal of particles from wastewater that are larger than about 3 μm. Typically used downstream of sedimentation (conventional treatment) or following coagulation/flocculation
Biological treatment		
Aerobic biological treatment	Biological metabolism of wastewater by micro-organisms in an aeration basin or biofilm (trickling filter) process	Removal of dissolved and suspended organic matter from wastewater
Oxidation pond	Ponds with 2–3 feet of water depth for mixing and sunlight penetration	Reduction of suspended solids, BOD, pathogenic bacteria, and ammonia in wastewater
Biological nutrient removal	Combination of aerobic, anoxic, and anaerobic processes to optimize conversion of organic and ammonia nitrogen to molecular nitrogen (N_2) and removal of phosphorus	Reduction of nutrient content of reclaimed wastewater
Disinfection	The inactivation of pathogenic organisms using oxidizing chemicals, ultraviolet light, caustic chemicals, heat, or physical separation processes (e.g. membranes)	Protection of public health by removal of pathogenic organisms

Table 6.2 (cont.)

Process	Description	Application
Advanced treatment		
Activated carbon	Process by which contaminants are physically adsorbed onto the surface of activated carbon	Removal of hydrophobic organic compounds
Air stripping	Transfer of ammonia and other volatile constituents from water to air	Removal of ammonia nitrogen and some volatile organics from wastewater
Ion exchange	Exchange of ions between an exchange resin and water using a flow-through reactor	Effective for removal of cations such as calcium, magnesium, iron, and ammonium, and anions such as nitrate
Chemical coagulation and precipitation	Use of aluminium or iron salts, polyelectrolytes, and/or ozone to promote destabilization of colloidal particles from reclaimed wastewater and precipitation of phosphorus	Formation of phosphorus precipitates and flocculation of particles for removal by sedimentation and filtration
Lime treatment	The use of lime to precipitate cations and metals from solution	Used to reduce scale-forming potential of water, precipitate phosphorus, and modify pH
Membrane filtration	Microfiltration, nanofiltration, and ultrafiltration	Removal of particles and micro-organisms from water
Reverse osmosis	Membrane system to separate ions from solution based on reversing osmotic pressure differentials	Removal of dissolved salts and minerals from solution; also effective for pathogen removal

waters. By reducing the quantity of treated wastewater discharged to surface waters, effluent requirements tend to be more favourable, particularly with respect to nutrient removal. Thus, water reuse is becoming an economic alternative for treatment facilities discharging into ecologically sensitive streams and estuaries.

Table 6.3 **Categories of municipal wastewater reuse**

Category of wastewater reuse	Treatment goals	Example applications
Urban use		
Unrestricted	Secondary, filtration, disinfection BOD_5: ≤ 10 mg/litre; turbidity: ≤ 2 NTU Faecal coliform: ND^a/100 ml Cl_2 residual: 1 mg/litre; pH 6–9	Landscape irrigation: parks, playgrounds, school yards Fire protection; construction Ornamental fountains; impoundments; In-building uses: toilet flushing, air conditioning
Restricted access irrigation	Secondary and disinfection BOD_5: ≤ 30 mg/litre; TSS: ≤ 30 mg/litre Faecal coliform: ≤ 200/100 ml Cl_2 residual: 1 mg/litre; pH 6–9	Irrigation of areas where public access is infrequent and controlled Golf courses; cemeteries; Residential; green belts
Agricultural irrigation		
Food crops	Secondary, filtration, disinfection BOD_5: ≤ 10 mg/litre; turbidity: ≤ 2 NTU Faecal coliform: ND/100 ml Cl_2 residual: 1 mg/litre; pH 6–9	Crops grown for human consumption and consumed uncooked
Non-food crops and food crops consumed after processing	Secondary, disinfection BOD_5: ≤ 30 mg/litre; TSS: ≤ 30 mg/litre Faecal coliform: ≤ 200/100 ml Cl_2 residual: 1 mg/litre; pH 6–9	Fodder, fibre, seed crops, pastures, commercial nurseries, sod farms, commercial aquaculture
Recreational use		
Unrestricted	Secondary, filtration, disinfection BOD_5: ≤ 10 mg/litre; turbidity: ≤ 2 NTU Faecal coliform: ND/100 ml Cl_2 residual: 1 mg/litre; pH 6–9	No limitations on body contact: lakes and ponds used for swimming, snowmaking

Table 6.3 (cont.)

Category of wastewater reuse	Treatment goals	Example applications
Restricted	Secondary, disinfection BOD_5: ≤30 mg/litre; TSS: ≤30 mg/litre Faecal coliform: ≤200/ 100 ml Cl_2 residual: 1 mg/litre; pH 6–9	Fishing, boating, and other non-contact recreational activities
Environmental enhancement	Site-specific treatment levels comparable to unrestricted urban uses Dissolved oxygen; pH Coliform organisms; nutrients	Use of reclaimed wastewater to create artificial wetlands, enhance natural wetlands, and sustain stream flows
Groundwater recharge	Site specific	Groundwater replenishment Salt water intrusion control Subsidence control
Industrial reuse	Secondary and disinfection BOD_5: ≤30 mg/litre; TSS: ≤30 mg/litre Faecal coliform: ≤200/ 100 ml	Cooling system make-up water, process waters, boiler feed water, construction activities and washdown waters
Potable reuse	Safe Drinking Water Requirements	Blending with municipal water supply Pipe to pipe supply

Source: Adapted from US Environmental Protection Agency (1992).
a. ND = not detected.

The increased implementation of wastewater reuse projects in various regions has facilitated the evolution of new water reuse alternatives. As treatment systems and applications are tested and design parameters are developed, technical barriers to wastewater reuse are reduced. Geographical, climatic, and economic factors dictate the degree and form of wastewater reclamation and reuse in different regions. In agricultural regions, irrigation is a dominant reuse application. In arid regions, such as California and Arizona, groundwater recharge is a major reuse objective either to replenish existing groundwater resources or to mitigate salt water intrusion in coastal areas. Industrial reuse of water varies with industries and locations. In contrast to the arid or semi-arid regions of the world where irrigation comprises a major beneficial use of reclaimed wastewater, wastewater reuse in Japan is dominated by non-potable urban uses such as toilet flushing, industrial use, and stream restoration and flow augmentation.

In Japan, several factors have contributed to increased implementation of wastewater reclamation and reuse. The vulnerability of the freshwater supply during drought years or in the aftermath of earthquakes or other catastrophic events has become evident in recent years. To increase the dependability and capacity of water resources throughout Japan, the government instituted a multi-faceted programme including the development of new water supply reservoirs, implementation of water conservation measures in large metropolitan areas, and wastewater reclamation and reuse where practical.

A comparison of water reuse applications for two locations in the United States (Florida and California) and for Japan is given in figure 6.2. The specific applications reflect the water balance associated with each region and other local conditions. Agricultural and landscape irrigation, groundwater recharge, and industrial reuse account for the majority of wastewater reuse in the United States.

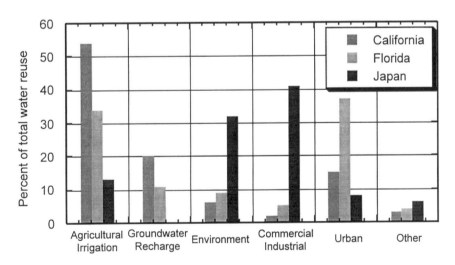

Fig. 6.2 **Comparison of the distribution of reclaimed water applications in California, Florida, and Japan (Source: adapted from Asano et al., 1996; Wright and Missimer, 1995)**

Health and regulatory requirements

The potential health risks associated with wastewater reclamation and reuse are related to the extent of direct exposure to the reclaimed water and the adequacy, effectiveness, and reliability of the treatment system. The goal of each water reuse project is to protect public health without unnecessarily discouraging wastewater reclamation and reuse. Regulatory approaches stipulate water quality standards in conjunction with requirements for treatment, sampling, and monitoring. To minimize health risks and aesthetic problems, tight controls are imposed on the delivery and use of reclaimed water after it leaves the wastewater treatment facility. Since major issues surrounding wastewater reclamation and reuse are often related to possible disease transmission, considerable research efforts have been directed towards protection of public health.

The specific criteria for wastewater reclamation in the United States are developed by individual states, often in conjunction with regulations on land treatment and disposal of wastewater (US Environmental Protection Agency, 1992). Some of the major differences among the approaches taken by individual states are associated with the degree of specificity provided in the rules. Also, some discrepancies exist from place to place in terms of monitoring and treatment requirements. For example, the state of California bases microbial water quality assessment on total coliform, whereas many other states require faecal coliform testing. The state of Arizona's wastewater reuse regulations contain enteric virus limits for the most stringent reclaimed water applications such as spray irrigation of food crops. The state of Florida requires monitoring of total suspended solids, instead of turbidity. California, Idaho, and Colorado criteria all stipulate requirements for wastewater treatment trains that include secondary treatment followed by coagulation, flocculation, clarification, filtration, and disinfection, or equivalent treatment, whereas other states have criteria that are less specific. Arizona, California, Florida, and Texas have comprehensive requirements according to the end use of the water (US Environmental Protection Agency, 1992).

In developing countries, the water quality criteria for using reclaimed water reflect a complex balance between protection of public health and the limited financial resources available for public works and other health delivery systems. In many cases, engineered sewage collection systems and wastewater treatment are limited, and reclaimed wastewater often provides an essential water resource and fertilizer source, as in the Valley of Mexico. A major concern for the use of wastewater for irrigation is control of enteric helminths such as hookworm, *Ascaris*, *Trichuris*, and, in certain circumstances, the beef tapeworm. In this context, it is necessary to provide protection from exposure to pathogens as a result of consuming crops irrigated with untreated or partially treated wastewater.

The degree of treatment required and the extent of monitoring necessary depend on the specific application. In general, irrigation systems are categorized according to the potential degree of human exposure. A higher degree of treatment is required for the irrigation of crops that are consumed uncooked, or the use of reclaimed water for irrigation of locations that are likely to have frequent human contact. To illustrate alternative regulatory practices governing the use of reclaimed wastewater for irrigation, the major microbiological quality guidelines from the World Health Organization (World Health Organization, 1989) and the state of California's current *Wastewater Reclamation Criteria* (State of California, 1978) are compared in table 6.4.

The WHO guidelines, which may be applicable to developing countries, emphasize that a series of stabilization ponds is necessary to meet microbial water quality requirements. In contrast, the California criteria stipulate conventional biological wastewater treatment followed by tertiary treatment including filtration and chlorine disinfection to produce effluent that is virtually pathogen free. Microbiological monitoring requirements also vary. The WHO guidelines require monitoring of intestinal nematodes, whereas the California criteria rely on treatment systems and monitoring of the total coliform density for assessment of microbiological quality.

The cost of wastewater reuse

Reported wastewater reclamation costs range widely from US$0.03 to US$0.65 per m^3 in the total cost in the United States. It is therefore important in comparing costs that differences in assumptions and factors associated with the allocation of costs to wastewater reclamation and reuse be correctly understood. Although costs associated with secondary treatment of wastewater are often considered to be water pollution control costs, they serve as a baseline cost for comparison with tertiary and/or advanced treatment facilities. A construction cost breakdown for various treatment processes within a secondary treatment system is estimated on the basis of 3,785 m^3/day,

Table 6.4 **Comparison of the World Health Organization's microbiological quality guidelines and criteria for irrigation and the state of California's current wastewater reclamation criteria**

Category	Reuse conditions	Intestinal nematodes[a]	Faecal or total coliforms[b]	Wastewater treatment requirements
WHO	Irrigation of crops likely to be eaten uncooked, sports fields, public parks	<1/litre	<1,000/100 ml	A series of stabilization ponds or equivalent treatment
WHO	Landscape irrigation where there is public access, such as hotels	<1/litre	<200/100 ml	Secondary treatment followed by disinfection
Calif.	Spray and surface irrigation of food crops, high exposure landscape irrigation such as parks	No standard recommended	<2.2/100 ml	Secondary treatment followed by filtration and disinfection
WHO	Irrigation of cereal crops, industrial crops, fodder crops, pasture, and trees	<1/litre	No standard recommended	Stabilization ponds with 8–10-day retention or equivalent removal
Calif.	Irrigation of pasture for milking animals, landscape impoundment	No standard recommended	<23/100 ml	Secondary treatment followed by disinfection

Sources: World Health Organization (1989); State of California (1978).

a. Intestinal nematodes (*Ascaris* and *Trichuris* species and hookworms) are expressed as the arithmetic mean number of eggs per litre during the irrigation period.

b. California Wastewater Reclamation Criteria are expressed as the median number of total coliforms per 100 ml, as determined from the bacteriological results of the previous seven days for which analyses have been completed.

whose total capital cost amounted to approximately US$0.50/m^3: primary treatment, 24 per cent; secondary treatment, 40 per cent; sludge treatment, 22 per cent; and control, laboratory, and maintenance buildings, 14 per cent.

To estimate the cost of a tertiary treatment system, several sources were used from the published literature (Sanitation Districts of Los Angeles County, 1977; Dames & Moore, 1978; Engineering-Science,

Table 6.5 **Comparison of tertiary treatment costs for water reuse**

	Life-cycle costs[a]	
Data source and plant flow	Title 22 US$/m^3	Direct filtration US$/m^3
Pomona virus study[b] 3,785 m^3/day	0.16	0.09
EPA cost estimation method[c] 3,785 m^3/day	0.23	0.15
City of Santa Barbara, CA[d] 5,262 m^3/day[e] 17,034 m^3/day[f]	– –	0.21 0.10
South Coast County Water District, CA[d] 2,915 m^3/day[g] 9,880 m^3/day[h]	– –	0.42 0.14
Irvine Ranch Water District 56,781 m^3/day	–	0.09
Los Angeles County Sanitation Districts (average of 4 treatment plants) 96,528 m^3/day	–	0.03
Monterey Regional Water Pollution Control Agency[i] 113,562 m^3/day	0.16	0.06

a. Costs adjusted to March 1993, using the Engineering News-Record Construction Cost Index (ENR CCI) of 5,106 for 20 US cities average. Reported costs include capital and O&M costs for tertiary processes only.
b. Costs include design and contingencies.
c. Costs include facilities planning, design, administrative and legal costs.
d. Costs include design, administrative, legal, and contingencies costs.
e. Cost for seasonal operation for landscape irrigation at 5,262 m^3/day average annual flow.
f. Cost for continuous operation at the design capacity of 17,034 m^3/day.
g. Cost for seasonal operation for landscape irrigation at 2,915 m^3/day.
h. Cost for continuous operation at the design capacity of 9,880 m^3/day.
i. Cost estimates based on Engineering-Science (1987) adjusted to October 1990 dollars.

1987; Young et al., 1988). Costs for both the "Title 22" process and the direct filtration process were estimated and are shown in table 6.5. The cost breakdown in one instance indicated that incremental tertiary treatment costs (chemical addition, filtration, solids treatment) were only US$0.06/m^3 while distribution costs, administrative charges (accounting, monitoring, overheads), and replacement reserve fees were US$0.12/m^3, US$0.04/m^3, and US$0.04/m^3, respectively. The critical importance represented by labour and energy costs in the water reuse system is noted (Young et al., 1988). The ratios of tertiary treatment costs for the "Title 22" treatment train to the direct filtration train range from 2.0 to 2.4 for capital costs, 3.9 to 5.6 for operation and maintenance (O&M) costs, and 2.4 to 2.9 for life-cycle costs for treatment capacities ranging from 3,785 m^3/day to 37,854 m^3/day (Richard et al., 1990).

However, there is a danger in comparing cost data from different studies and locations because of differing underlying assumptions, which often are not explicitly stated. As seen from the data in table 6.5, costs are significantly affected by the fraction of utilization of a facility over the course of a year. Economic assumptions about useful life and interest rates affect the amortization of capital costs embedded in unit costs. Reported costs may represent current expenses for old facilities and not reflect costs to construct those facilities at today's prices, as seen in Irvine Ranch Water District in California.

One factor that appears to affect costs significantly is the degree of utilization of available capacity in the treatment plant. Maximum utilization can be achieved by: (1) seasonal storage of effluent to compensate for seasonal slack in water reuse demand, (2) obtaining a mix of reclaimed water uses to reduce seasonal demand, or (3) using alternative water supplies to meet peak demand.

The future of water reuse

Significant progress has been made with respect to developing sound technical approaches to producing a high-quality and reliable water source from reclaimed wastewater. Continued research and demonstration efforts will result in additional progress in the development of water reuse applications. Some key topics include: assessment of health risks associated with trace contaminants in reclaimed water; improved monitoring techniques to evaluate microbiological quality; optimization of treatment trains; improved removal of wastewater particles to increase disinfection effectiveness; the application of membrane processes in the production of reclaimed water; the effect of reclaimed water storage systems on water quality; evaluation of the fate of microbiological, chemical, and organic contaminants in reclaimed water; and the long-term sustainability of soil and aquifer treatment systems.

A key to promoting the implementation of water reuse is the continued development of cost-effective treatment systems. One such system is a series of smaller satellite treatment systems upstream from the major regional treatment facilities. Emerging advanced treatment technologies such as membrane systems can be used to produce reclaimed water of almost any quality to fit the required application.

To date, however, the major emphasis in wastewater reclamation and reuse has been on non-potable applications such as agricultural and landscape irrigation, industrial cooling, and in-building applications such as toilet flushing. Although direct potable reuse of reclaimed municipal wastewater is, at present, limited to extreme situations, it has been argued that the practice is no different from withdrawing raw water from a polluted water source for the domestic water supply. It is further argued that there should be a single water quality standard for potable water. If reclaimed water can meet this standard, it should be acceptable regardless of the source of the water. Although indirect potable reuse through groundwater recharge or surface water augmentation has gained support, some concerns still remain regarding trace organics, treatment and reuse reliability, and, particularly, public acceptance.

Considering that modern wastewater reclamation and reuse started in the 1960s, remarkable progress has been made. A cautious and judicious approach is warranted to avoid potential health consequences if a wastewater reuse project is not successful. In addition, the importance of public confidence cannot be underestimated. It is clear that reclaimed wastewater is a viable water resource. Continued research and development will provide a sound scientific basis for crossing the threshold to direct potable reuse, when necessary. As technology continues to advance and the reliability of wastewater reuse systems is demonstrated, applications for wastewater reclamation, recycling, and reuse will continue to expand and become a vital element in integrated water resources management in mega-cities in the twenty-first century.

Summary and conclusions

The successful implementation of wastewater reuse options in a water resources management programme requires careful planning, economic and financial analyses, and the effective design, operation, and management of wastewater reclamation, storage, and distribution facilities. Technologies for wastewater reclamation and purification have developed to the point where it is technically feasible to produce water of almost any quality, and advances continue to be made. Current water reclamation strategies incorporate multiple measures to minimize the health and environmental risks associated with various reuse applications.

A combination of source control, advanced treatment process flow schemes, and other engineering controls provides a sound basis for increased implementation of water reuse applications. The feasibility of producing reclaimed water of a specified quality to fulfil multiple water use objectives is now a reality owing to the progressive evolution of technologies and the understanding of health and environmental risks. However, the ultimate decision to harvest reclaimed wastewater is dependent on economic, regulatory, and public policy factors reflecting the demand and need for a dependable water supply and water pollution control facing the mega-cities.

Through integrated water reuse planning, as discussed in this chapter, the use of reclaimed wastewater will provide sufficient flexibility to allow a water agency to satisfy short-term needs as well as to increase water supply reliability. With an increasing emphasis on the planning and implementation of wastewater reclamation and reuse facilities, accurate cost data are essential. Thus, cost information was presented, although there are significant variations in wastewater reclamation and reuse costs. Although droughts often underscore the need for wastewater reclamation and reuse, wastewater reuse is by no means a water resources management alternative for drought years only. It should be considered an integral and permanent part of water resources planning in the future.

References

Asano, T. (ed.). 1998. *Wastewater Reclamation and Reuse*, Water Quality Management Library Vol. 10, Technomic Publishing Co., Lancaster, PA.

Asano, T. and Levine, A. D. 1995. "Wastewater Reuse: A Valuable Link in Water Resources Management," *Water Quality International*, No. 4, pp. 20–24.

Asano, T., Maeda, M., and Takaki, M. 1996. "Wastewater Reclamation and Reuse in Japan: Overview and Implementation Examples," *Water Science and Technology*, Vol. 34, No. 11, pp. 219–226.

Dames & Moore, Water Pollution Control Engineering Services. 1978. *Construction Costs for Municipal Wastewater Treatment Plants: 1972–1977*, EPA 430/9-77-013, Office of Water Program Operations, US Environmental Protection Agency, Washington, D.C.

Dryden, F. D., Chen, C.-L., and Selna, M. W. 1979. "Virus Removal in Advanced Wastewater Treatment Systems," *Journal Water Pollution Control Fed.*, Vol. 51, No. 8, p. 2098.

Engineering-Science. 1987. *Monterey Wastewater Reclamation Study for Agriculture, Final Report*, prepared for Monterey Regional Water Pollution Agency, Pacific Grove, CA.

Lindh, G. 1992. "Urban Water Studies," in *Water, Development and the Environment*, edited by W. James and J. Niemczynowicz, Lewis Publishers, Chelsea, MI, pp. 23–24.

National Research Council. 1993. *Managing Wastewater in Coastal Urban Areas*, National Academy Press, Washington, D.C.

Richard, D., Crites, R., Tchobanoglous, G., and Asano, T. 1990. "The Cost of Water Reclamation in California," presented at the 62nd Annual Conference of the California Water Pollution Control Association, South Lake Tahoe, CA.

Sanitation Districts of Los Angeles County. 1977. *Pomona Virus Study – Final Report*.

Sheikh, B., Cort, R. P., Kirkpatrick, W. R., Jaques, R. S., and Asano, T. 1990. "Monterey Wastewater Reclamation Study for Agriculture," *Research Journal Water Pollution Control Federation*, Vol. 62, No. 3, pp. 216–226.

State of California. 1978. *Wastewater Reclamation Criteria. An Excerpt from the California Code of Regulations, Title 22, Division 4*, Environmental Health, Department of Health Services, Berkeley, CA.

——— 1990. *California Municipal Wastewater Reclamation in 1987*, State Water Resources Control Board, Office of Water Recycling, Sacramento, CA, June.

US Environmental Protection Agency. 1992. *Guidelines for Water Reuse Manual*, EPA/625/R-92/004, Washington, D.C.

World Health Organization. 1989. *Health Guidelines for the Use of Wastewater in Agriculture and Aquaculture*, Report of a WHO Scientific Group, Geneva, Switzerland.

Wright, R. R., and Missimer, T. M. 1995. "Reuse. The U.S. Experience and Trend Direction," *Desalination and Water Reuse*, Vol. 5, No. 3, pp. 28–34.

Young, R. E., Lewinger, K., and Zenk, R. 1988. "Wastewater Reclamation – Is It Cost Effective? Irvine Ranch Water District – A Case Study," Proceedings of Water Reuse Symposium IV, Implementing Water Reuse, American Water Works Association Research, Denver, CO, pp. 55–64.

7

The role of the private sector in the provision of water and wastewater services in urban areas

Walter Stottmann

Introduction

The present situation of the water and sanitation sector

Governments around the world are facing a great challenge in improving access to good-quality, reasonably priced water and sanitation services for their citizens. Today, over 1 billion people are without safe water, almost 2 billion lack adequate sanitation, and the waste of over 4 billion is discharged with inadequate or no treatment. The most severe problems exist in low-income countries, where only about 60 per cent of the population has access to a public water supply and only 40 per cent to adequate sanitation. Even those with access suffer from poor and unreliable service. The huge task of treating municipal wastewater has barely begun in middle- and low-income countries. The present crisis is exacerbated by the fact that large portions of the existing water and sanitation infrastructure need to be rehabilitated or replaced. As urban populations continue to grow relentlessly, particularly in low-income countries, water and wastewater infrastructure has to be provided to an ever-growing number of people. Population forecasts indicate that over the next 25 years the world's population will increase in the order of 40–50 per cent. That means that by 2025 an additional 2.5 billion people will need to be provided with water and sanitation services. Much of these investments will focus on providing better services in the world's rapidly growing large metropolitan conglomerations. By the year 2010 there will be more than 30 mega-cities with more than 10 million inhabitants. The challenge is indeed staggering. It is estimated that low- and medium-income countries alone may have to spend up to 1 per cent of their gross domestic product to rise to the challenge. This translates into resource needs in the order of US$50 billion per year.

Mobilizing these resources and investing them cost-effectively will require fundamental changes to current sector development policies and the way water and wastewater infrastructure is managed. In particular it will require policy changes, incentives, and extensive capacity-building to encourage the emergence of better-managed and more efficient sectoral institutions on all levels capable of operating systems more effectively and making more cost-effective investment decisions. All too many water and wastewater institutions today run systems in a less than optimal way. A World Bank review of more than 120 public sector water projects completed over the past 20 years found that only some 3 per cent of the participating water and wastewater utilities were operating at acceptable levels of performance. Unaccounted for water, an indicator of system leakage and unauthorized or unrecorded withdrawal of water, was found to be in the range of 40–60 per cent compared with an acceptable industry standard of 10–20 per cent. It is estimated that operational inefficiencies of utilities in low- and middle-income countries may add cost on the order of US$10 billion per year. Clearly, demand management to reduce consumption and conserve precious and dwindling water resources and efficiency improvements to combat the waste of water by leakage and to make networks and plants work better while consuming fewer chemicals and less energy will be the most important tasks in the future. Capital investments must be planned more carefully than in the past. Too often new investments stress the creation of new capacity, whereas rehabilitating and making existing systems more efficient might at least initially provide more cost-effective solutions.

Technical inefficiencies are mirrored by poor management and low productivity at the great majority of municipal water and wastewater utilities. Public utilities tend to be severely overstaffed, with five to seven times as many employees per connection than best industry practice. Antiquated management techniques and administrative systems run by often poorly trained and motivated managers and staff are problems common to most water utilities. Another major problem facing the sector is restrictive pricing policies defended by governments interested in short-term political gain rather than facing up to the long-term challenge of providing adequate services to their people. Massive under-pricing and failure to collect tariffs effectively do not allow many utilities to raise the financial resources needed for maintaining and operating water and wastewater systems adequately and prevent the generation of much-needed funds for investing in system rehabilitation and expansion. At the same time, severe fiscal constraints and budgetary pressures in most countries place limits on the contributions that public budgets can make for the water and wastewater sector. Without revising ill-considered and short-sighted cost recovery policies, many countries will not be able to raise enough resources for adequate maintenance and operation of existing assets and certainly not for the immense investment needs of the future. At the same time, improving the human capacity and techniques to manage water and wastewater utilities better, operate networks and plants more efficiently, and plan and select investments more cost-effectively will have to be a most important part of the future development effort to ensure that the scarce resources that can be generated for the sector will be employed as efficiently as possible. There is wide agreement in the international development community that tackling the enormous and ever-growing need for improved and expanded services in many countries will require a radical rethinking of present sector development policies. Old approaches and policies based on the public sector alone will not be able to provide the know-how or the resources to address the urgent problems facing the sector.

The private sector – part of the solution

Realizing the limits on the capacity of the public sector, governments throughout the world, including those in the developing countries of Asia, Africa, and Latin America, are increasingly looking towards the private sector to leverage the financial and managerial capacities of the state. Over the past decade, several large cities in low- and middle-income countries have decided to turn to the private sector. Prominent examples include Buenos Aires in Argentina; Abidjan in Côte d'Ivoire; Manila in the Philippines; Dakar in Senegal; Djakarta in Indonesia; and other secondary cities in Côte d'Ivoire and Senegal. These cities turned to the private sector with one or more of the following objectives in mind: (1) to acquire technical and managerial expertise and better technologies to improve economic efficiency in both operating performance and the use of capital investment; (2) to inject investment capital into the sector or gain access to private capital markets; (3) to insulate the sector from short-term political intervention in utility operations and limit opportunities for intervention by powerful interest groups; (4) to turn around or restructure failing public enterprises

The role of the private sector in the provision of water and wastewater theories is not new. In market economies, the private sector traditionally has played an important role as contractor or consultant and supplier of equipment and services. Less frequent, although not at all uncommon, is the direct involvement of the private sector in the management and operation of water and wastewater utilities and in providing finance for investments. The private sector has been a partner for decades in many parts of the world. In the United States of America, some 20 per cent of the population is served by privately managed utilities. In Europe, the involvement of the private sector dates back to the nineteenth century. In France, about 75 per cent of all urban water connections are operated by private firms under management contracts, leases, or concessions. Even in some low- and middle-income countries, the notion of private sector involvement has been there for quite some time. In a number of cities – Barranquilla in Colombia, and Alexandria in Egypt, for example – water and wastewater utilities were initially founded in the 1920s and 1930s as private enterprises, only to be nationalized and turned into public enterprises in the 1950s.

Over the past decade, multilateral development institutions concerned with the development of public infrastructure such as the World Bank, its affiliate the International Finance Corporation, and the European Bank for Reconstruction and Development (EBRD) have adopted policies encouraging private involvement in the sector. This policy shift has its origin in the generally disappointing performance of the public sector and mounting evidence that the private sector can indeed be instrumental in helping the sector develop more efficiently in the future. A recent World Bank study confirms that, in a number of projects, private enterprise has been associated with substantial benefits to consumers in terms of expanded coverage and quality of service as well as significant improvements in operational

159

efficiency. In fact many practitioners in the development community have come to the conclusion that the challenge of the future cannot be tackled without the involvement of the private sector.

In spite of the recent growth of private sector participation in the sector, decision-makers in many countries remain highly sceptical about private involvement. This reluctance may have its origin in historical traditions, as is clearly the case in the transition countries of Central and Eastern Europe. In other countries, decision-makers find it difficult to reconcile private sector involvement with the long-held notion of water as a social good. In other countries, political leaders accustomed to use public enterprises for their own political purposes are not willing to relinquish control. Perhaps the most common denominator for the rejection of private involvement is lack of knowledge and misunderstanding about the potential role of the private sector and how public and private partners can build partnerships that are beneficial to both. The purpose of this paper is to rectify some of the misconceptions of the role of private enterprise in the sector and outline the conditions under which the private sector could play an instrumental part in the development of the water and wastewater sector. It will describe the various ways in which the private sector can get involved and how to go about planning and forming lasting public/private partnerships.

Options for private participation in municipal water and wastewater

The involvement of the private sector in municipal water and sanitation can take many different forms. The main options for private sector participation can be distinguished by how they allocate responsibility for such functions as asset ownership and capital investment between the public and private sectors (table 7.1). They are distinguished in the assignment of risks and responsibilities between the public, represented by government, and the private operator or investor. A description of the main characteristics of these options follows. Table 7.2 gives some examples of the various types of private sector participation in place in water and sanitation.

Service contracts

Under service contracts, a private sector firm is contracted to perform specific discrete tasks, for example installing or reading meters,

Table 7.1 **Allocation of key responsibilities for private participation options**

Option	Asset ownership	Operations and maintenance	Capital investment	Commercial risk	Contract duration
Service contract	Public	Public and private	Public	Public	1–2 years
Management contract	Public	Private	Public	Public	3–5 years
Lease	Public	Private	Public	Shared	8–15 years
Concession	Public	Private	Private	Private	25–30 years
BOT/BOO	Private and public	Private	Private	Private	20–30 years
Divestiture	Private, or private and public	Private	Private	Private	Indefinite (may be limited by licence)

repairing pipes, or collecting accounts. They are contracted for short periods, typically from six months to two years. The main benefit of service contracts is that they allow the public utility to tap private sector expertise for specific tasks and open these tasks to competition. Service contracts are widely used. In India, for example, the Madras Water Enterprise has contracted services ranging from the management of its vehicle fleet to the operation and maintenance of sewage pumping stations. The water utility in Santiago in Chile has contracted out services accounting for about half its operating bud-

Table 7.2 **Some private sector contracts in place in water and sanitation**

Option	Water	Sanitation	Water and sanitation
Management or service contract	Colombia, Gaza, Malaysia, Turkey	United States	Puerto Rico, Trinidad and Tobago
Lease	France, Guinea, Italy, Senegal, Spain, Germany		Czech Republic, France, Poland
Concession	Côte d'Ivoire, France, Macao, Malaysia, Spain	Malaysia	Argentina, France
BOT	Australia, China, Malaysia, Thailand	Chile, Mexico, New Zealand	
Divestiture	England and Wales		England and Wales

161

get, including computer management, engineering, and repair, maintenance, and rehabilitation of the water and sewerage network. To enhance competition, the Santiago utility has at least two service contracts for each kind of task.

Although relatively simple, service contracts must be carefully specified and monitored. If a utility is poorly managed, its service contracts probably will be too. Service contracts are at best a cost-effective way to meet special technical needs for a utility that is already well managed and commercially viable. They cannot substitute for reform in a utility plagued by inefficient management and poor cost recovery.

Management contracts

Management contracts transfer responsibility for the operation and maintenance of the entire water and/or wastewater systems to a private sector operator for a period of generally three to five years. In its simplest form, the management contract provides for a fixed fee for reimbursing the contractor for its services. More sophisticated management contracts introduce incentives for efficiency, by defining performance targets and basing remuneration at least in part on meeting these targets. Specifying clear and indisputable targets, however, is often difficult, especially when information about a system's current performance is limited. Because management contracts leave all responsibility for investment with the government, they are not a good option if a government seeks to access finance for new investments.

Management contracts are most useful where the main objective is rapidly to enhance a utility's technical capacity and efficiency and strengthen its management and operation. They are a good first step toward more full-fledged private sector involvement where conditions make it difficult for the government to commit to a long-term arrangement initially or the private sector is not prepared to undertake capital investment or accept commercial or political risk. A management contract might be the best arrangement in a city where tariff income is too low or too uncertain to support a private contract on tariff revenue. The management contract is an ideal initial short-term arrangement that can lead to immediate efficiency improvements while giving the government the time to improve the political, technical, and financial base of the utility for a more advanced private sector arrangement such as a lease or concession. This was the approach adopted in Mexico City and in Trinidad and Tobago, where management contracts were signed with the understanding that more advanced arrangements were to follow in the future.

Leases

Under a lease arrangement, a private firm leases the assets of a utility from the government and takes on the responsibility for operating and maintaining them. Because the private firm, as lessor, in effect buys the rights to the income stream from the utility's operations, minus the lease payment, it assumes much of the commercial risk of the operations. Under a well-structured contract, the lessor's profitability will depend on how much it can reduce costs while still meeting the quality standards set out in the lease contract. A lease thus has built-in incentives for improving operating efficiency. Leases leave the responsibility for financing and planning investments with the government. If major new investments are needed, the government must raise the finance and coordinate its investment programme with the lessor's operational and commercial programme.

Leases are most appropriate where there is scope for large gains in operating efficiency but only limited need or scope for new investments. Leases have also sometimes been advocated as stepping stones toward more full-fledged private sector involvement – a concession, for example. Their administrative complexity and their need for sustained government commitment are nearly as demanding as for more advanced private participation options – concessions, for example. "Pure" leases are rare, however. Most place some responsibility for investment on the private partner, if only for rehabilitation works. These contracts operate as a hybrid between a lease and a concession contract. Leases have been widely used in France and Spain and are currently in place in the Czech Republic, Guinea, and Senegal. They were also used in the Côte d'Ivoire until replaced by a concession.

Concessions

A concession gives the private partner responsibility not only for the operation and maintenance of a utility's assets but also for investments. Asset ownership remains with the government, however, and full use rights to all the assets, including those created by the private partner, revert to the government when the contract ends – usually after 25–30 years. Concessions are usually bid by price: the bidder that proposes to operate the utility and meet the investment targets for the lowest tariff wins the concession. The concession is governed by a contract that sets out such conditions as: performance targets (service coverage, quality), efficiency standards, arrangements for capital investment, mechanisms for adjusting tariffs, and arrangements for arbitrating disputes.

The main advantage of a concession is that it passes full responsibility for operations and investment to the private sector and so brings to bear incentives for efficiency in all aspects of utility management and operation. A concession is, therefore, an attractive option where large investments are needed to expand the coverage or improve the quality of services. From the government's perspective, administering a concession is a complex business, however, because it confers a long-term monopoly on the concessionaire. Careful oversight is therefore important in determining the success of the concession, particularly the distribution of its benefits between the concessionaire (in profits) and consumers (in lower prices and better service). Concessions have a long history of use in infrastructure in France. Recently they have spread to the developing world, where they have been used, for example, in Buenos Aires in Argentina, Macao, Manila in the Philippines, and Malaysia.

Joint venture leases and concessions

In several countries, most notably in Spain, a specific form of leases and concessions has evolved that involves the creation of public/private joint venture firms that then take on a lease or a concession. Under this arrangement, a company is formed with a state entity (a municipal government, for example) holding 51 per cent of the shares and the private operator or a financial institution (or both) holding the remaining shares. By limiting the private sector's control, these joint ventures can help secure stakeholders' agreement to private sector participation. Also, by demonstrating public commitment to the venture, they can reduce the private sector's perception of risk. The joint venture model, however, creates obvious conflicts of interest if municipal government is both the supervisor and the part owner of the joint venture utility-operating company. Another issue is the extent to which the private firm can exercise management control, especially if it has only a minority shareholding in the joint venture. Without such control, the private firm may not feel that its interests are protected and may not be able to produce the efficiency gains expected from private involvement. Most joint ventures address control issues through detailed clauses in the company's by-laws allowing both parties to vet managerial appointments and other key decisions. These clauses may foster partnership, but at the expense of "muddled" governance relationships. In Gdansk in Poland, where a mixed ownership lease contract is in operation, the relationship between the city and the project company has been described as tense and complex. The terms of the lease have been renegotiated four times in five years and tariff increases have lagged behind inflation. In spite of these problems, the Gdansk joint venture operator has been widely successful in improving service quality and system efficiencies in the city of Gdansk.

Build–operate–transfer contracts

Build–operate–transfer (BOT) arrangements resemble concessions but are normally restricted to large discrete projects, such as a water or wastewater treatment plant. In a typical BOT arrangement, a private firm constructs a new plant, operates it for a number of years, and at the end of the contract, 25–30 years later, turns it over to the public. The government or the distribution utility pays the BOT partner for water from the project, at a price calculated over the life of the contract to cover its construction and operating costs and provide a reasonable return. The contract between the BOT contractor and the utility is usually on a take-or-pay basis, obligating the utility to pay for a specified quantity of water whether or not that quantity is consumed. This places all demand risk on the utility. Alternatively, the utility might pay a capacity charge and a consumption charge, an arrangement that shares the demand risk between the utility and the BOT concessionaire. BOTs work well if the issue is to increase production capacity. BOTs for capacity expansions, however, do little to enhance utility management or improve operating efficiency. In fact, building a new water treatment plant may be a poor investment choice, independent of the financing arrangement, if the new production is wasted in a leaky distribution system.

There are many possible variations on the BOT model, including build–operate–own (BOO) arrangements, in which the assets remain indefinitely with the private partner, and design–build–operate (DBO) arrangements, in which the public and private sectors share responsibility for capital investments. BOTs may also be used for plants that need extensive overhauls – in arrangements sometimes referred to as ROTs (rehabilitate–operate–transfer). Examples of recent BOTs or similar arrangements are water treatment plants in Australia, Malaysia, and China and sewage treatment plants in Chile, New Zealand, and Slovenia.

Full or partial divestiture

Divestiture of water and sewerage assets – through a sale of assets or shares or through a management buyout – can be partial or complete. A complete divestiture, like a concession, gives the private sector full responsibility for operations, maintenance, and investment, but, unlike a concession, divestiture transfers ownership of the assets to the private sector. A concession assigns the government two primary tasks: to ensure that the utility's assets – which the government continues to own – are used well and returned in good condition at the end of the concession; and, through regulation, to protect consumers from monopolistic pricing and poor service. A divestiture leaves the government only the task of regulation, since, in theory, the private company should be concerned about maintaining its asset base.

Although widely used in other infrastructure sectors, private ownership of assets in the water and sanitation sector has been very limited. Some utilities in the United States own assets. On a large scale, full divestiture of water and wastewater assets was enacted in the 1980s in England and Wales. These assets were sold in the private market to private firms and investors, which now operate them under a regulatory regime provided by government.

Hybrids

In practice, private sector arrangements are often hybrids of the options outlined above. For example, leases often pass some responsibility for small-scale investment to the private sector, and management contracts may have revenue-sharing provisions that make them a little like leases. Options might also be used in combination – for example, a build–operate–transfer contract for bulk water supply might be combined with a management or lease contract for operating the distribution system.

Competition and regulation

Providing water and wastewater services is a business with natural monopoly features. Direct competition among suppliers to provide services to individual customers would require the construction of duplicate plants and networks, which is economically not justifiable. Unlike private enterprises that can declare bankruptcy and cease production, water and wastewater utilities cannot be allowed to stop operating because they provide a service essential to human well-being and health. As an essential service, it cannot be left in the hands of the private sector unsupervised. In fact, the natural monopoly features of the business oblige government, as the representative of consumers, to retain ultimate responsibility for the efficient provision of the service. Private participation in the management of the water and wastewater sector is, therefore, possible only in partnership with the government. Private participation does not mean that the public sector disengages entirely or relinquishes control; rather it signals a new division of labour with private sector partners based on the comparative advantage of each.

The challenge, then, is to structure relationships between the government/public and the private sector such that: (a) customers receive an efficient service of appropriate quality at a fair price; and (b) the private sector obtains a fair rate or return for providing the service. To achieve that, the private and public sectors must work together within a framework of checks and balances that on the one hand leaves the private partner free to manage its affairs as efficiently as possible, and on the other provides for sufficient control by the public sector to ensure that the private partner fulfils its obligations. In doing so, the public sector needs to resist the temptation of micro-managing the utility operator for short-term political gain, a practice that is at the root of the many failures in the public provision of services. Two key mechanisms are employed to guide the private/public partnership: competition and regulation.

Competition

In spite of the monopolistic features of the business, competition is a key element in the effort by government to ensure that the private sector discharges its responsibilities as efficiently as possible. The way competition is brought to bear may be divided into two broad concepts: competition *for the market* and competition *in the market*.

Competition for the market involves competition for the right to serve an entire segment of the market; for example, a management contract, a lease, or a concession covering an entire municipal area. It comes into play when a new contract is being considered or an existing one is up for renewal. The opportunity for competition for the market is greatest if the contracts are re-bid frequently. For simple service contracts, it may be feasible to hold annual or bi-annual bidding rounds to maintain competitive pressure for efficient performance. For lease contracts or BOTs with long contracting periods, the opportunity for re-bidding diminishes, which virtually eliminates the pressure of market competition.

Competition in the market refers to head-to-head competition among service providers. These can compete directly to serve each other's customers, as allowed in England and Wales. This may require arrangements for use of a service provider's networks by another. Melbourne, Australia, is a good example. There the water network was divided into four zones each managed by a different operator, with the possibility of one operator expanding into another's territory. Competition in the market also occurs when tanker operators compete with each other or with suppliers of piped water. Competition can also be introduced by unbundling, i.e. contracting out the main componenets of a water system – water production, treatment, transmission, and distribution – to different operators. Another way of bringing about competition occurs when regulators use comparisons of performance as a way of encouraging better performance. This approach, referred to as "yardstick" competition, entails, for example, awarding contracts for different geographical areas in one market. In Paris in France, for example, services are provided by two concessionaires each responsible for a separate area of the city. Another example is Manila in the Philippines, where the city's service area was bifurcated and separate contracts awarded for each half of the city. Such arrangements allow the direct comparison of performance between two private operators. Performance comparisons across several cities are an important tool to reduce the potential for abuse of monopoly powers. Confronted with good performance by other operators in a similar situation, private operators are forced to strive for efficient performance in an effort to keep up their reputations, because their ability to win contracts will depend in part on their performance elsewhere. Such reputation effects will work best where there are provisions for publishing consistent performance information. The success and fairness of "yardstick" competition depend of course on a degree of comparability across jurisdictions, which is not always possible. In England and Wales, for example, the government regulator had to realize early on that significant size and geographical differences between service areas make a fair comparison difficult.

The extent of competition in or for the market is of course very much dependent on the number of private firms capable of offering quality services to the market. Although their numbers have increased substantially over the past decade, there are no more than two dozen or so international firms that can or are willing to offer their services world-wide. Most of them are large international conglomerates concentrated in France and the United Kingdom. This shortage of qualified firms limits the possibility for competition except for very attractive projects. A call for a concession in a large city in a country with a private sector friendly environment will generate the interest of a great segment of the industry, whereas there may be little interest on the part of the private sector to get involved in a smaller city in a country with limited financial resources and a legal/political environment less friendly to the private sector. The prospects for a future greater role of the private sector in the water and wastewater field will depend on increasing the number of qualified firms willing to participate in the market. Of particular importance for the spread of private participation in the developing countries is the emergence of a capable domestic private sector.

Regulation

In a natural monopoly situation, competition alone cannot ensure that the private sector will perform efficiently. Governments will need to introduce safeguards against potential abuses of the private operator's monopoly position by providing incentives for the private sector to operate efficiently, and by carefully monitoring performance. There will always be a need for public oversight of the activities of the private operator, although the level of control and monitoring will vary depending on the private sector option chosen. The nature of the relationship between the public and private sector – and the rights, responsibilities, and risks it entails for each partner – are determined by the regulatory framework that government adopts. The need for regulation is least for simple operations and maintenance contracts. The operator receives a fixed fee for a specified task and the contract can be re-bid frequently. Long-term concessions and divestiture will require a much more comprehensive system of public scrutiny and vigilance. Figure 7.1 illustrates how the level of risk and

The role of the private sector

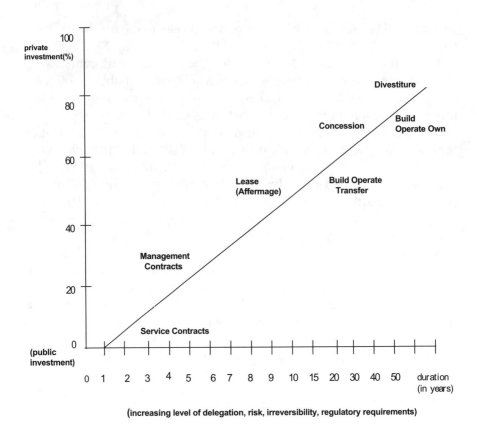

Fig. 7.1 **Range of options for private sector participation in the municipal water and wastewater sector**

regulation increases as the private sector takes on more and more responsibility and risk.

In most countries there are two distinct mechanisms that determine the regulatory framework under which a private contract is managed: (1) general and sector-specific laws that establish the broad principles of regulatory policy and set national standards, for example for minimum drinking water quality or effluent standards for wastewater discharges; and (2) contractual agreements between the private operator and government that govern the issues not covered by general or sector-specific law. Between the broad principles of general and sector law and the specifics set out in a contract there inevitably exists a grey area that needs to be covered through regulation. Sometimes it is argued that a tightly specified contract can remove the need for direct regulation. This is rarely the case. Even for a short-term management contract, government needs to monitor performance against the contract and adjust contract conditions for changing circumstances. For longer-term concession contracts or BOTs, it is usually neither possible nor desirable to have highly specific contracts. Detailed and rigid contract conditions do have advantages in that they leave very little room for dispute and minimize political discretion, but they also limit responses to changing social, economic, and technical conditions, which are bound to develop in unpredictable ways over the life of a 20-year contract. Because it is impossible to "get everything right" at the outset, specific and rigid contracts make it difficult to fine-tune or improve on the original arrangements. This can be particularly damaging in situations where the initial information is limited. Highly specific contracts lead to more frequent renegotiating. This tends to benefit the private contract holder, which is likely to have a stronger bargaining position and better information. Hence, a delicate balance needs to be struck between specific contract provisions, which reduce the regulator's role to one of monitoring compliance with a set of existing requirements, and more flexible arrangements, which allow regulatory discretion. This balance will of course be different for different private participation options and the capacity of the government to regulate.

Considerations in defining a regulatory framework

In defining a regulatory system, several important questions must be answered: What should be regulated and what left to the contract? Who should regulate? What discretion should a regulator have? How can it be ensured that the regulator is independent and accountable?

Areas to be regulated
The duties of the regulator will depend on the kind of private sector arrangement adopted, the degree to which service conditions and price adjustments are already specified in law or contract, and the existence of regulators concerned with the water and wastewater business. For example, many countries do have anti-monopoly agencies, river basin and environmental authorities, health and safety inspectorates, or utility commissions. In many cases, however, some regulatory capacity is generally required to deal with increases in pricing, to monitor private operator performance and contractual

compliance, to receive complaints and arbitrate disputes between a utility and its consumers, and to impose sanctions if agreed standards are not met.

Price regulation
A central task of regulation is usually to deal with prices and preventing hidden price rises through reduced standards of service. There are two basic types of price regulation: (a) rate of return or profit control, under which the regulator places limits on the returns on invested capital, dividends payable to shareholders, or capital reserves; and (b) price control – regulators peg allowable price increases to an independent measure such as the retail price index, possibly adjusted for expected efficiency gains.

Rate of return regulation is used with capital investments. After determining an appropriate rate of return on the investment, the regulatory agency sets the maximum return that the utility may earn on its assets for a specified period. The advantage of this approach is that it keeps prices at competitive levels and gives comfort to investors that they will be able to earn a return on their investment – which may lower the cost of capital. But in practice there are several problems. It can reduce the incentives of regulated utilities to lower maintenance and operating expenditures while over-investing in capital outlays. If the allowed rate of return is greater than the utility's cost of capital, the utility will be inclined to maximize its profits by substituting capital investments for other inputs to its production. And if the allowed rate of return is less than the cost of capital, the utility may have an incentive to use a less capital-intensive method of production than it otherwise would. In both cases the result will be higher costs than necessary.

Price control regulation involves the setting of a general "cap" on prices. This cap is usually determined by reference to the inflation rate and to an assessment of the potential for efficiency improvements by the regulated utility. The main advantage of this approach is that it provides utilities with an incentive to reduce costs and operate efficiently, because they keep any profits generated by increases in productivity above those postulated by the regulator. The approach has several drawbacks. If the price is set too high, the private trade operator or investor will earn high profits, which may be unacceptable to the public. If the price is set too low, the level and quality of service may fall because the operator finds it impossible to earn a reasonable rate of return; investors are then placed at risk and the cost of capital may increase accordingly. Expected productivity gains may also be set too high, a risk the investor must confront each time the cap is renegotiated, which may be five to six times over the life of a 30-year concession. Price caps may not be attractive if the primary concern is to promote new investments by the regulated utility. Price cap regulation is used in England and Wales, where a national agency reviews pricing policy and tariffs every five years. Both rate of return and price cap regulation require extensive and reliable information on all aspects of the utility business.

Locus of regulation
As the responsibility for the provision of water and wastewater services is being more and more decentralized throughout the world, municipal government increasingly assumes the responsibility for regulation. Decentralized regulation can generally be more responsive to local needs and conditions, ease monitoring, and ensure better access to information, but it can increase regulatory cost through replication of regulatory agencies, reduce regulatory effectiveness, and, because of lack of capacity, increase the danger of poor regulation. Where regulatory functions are decentralized, national governments can still put in place arrangements to support effective and consistent regulatory decisions. Options include: providing training for regulatory staff; publishing national performance indicators; creating a central or regional agency with auditing functions to monitor the effectiveness of local regulators and reduce the risk of regulatory capture; requiring local regulators to publish the results of their monitoring activities and regulatory decisions; providing reporting and monitoring guidelines to help ensure that utility performance is measured consistently and in a way that eases comparison; and requiring local regulators to employ professional independent monitors – private audit firms, for example. All of these measures leave regulatory authority at the local level, where it may best be located, but attempt to ensure that a higher level of government has a role of monitoring the performance of utilities and local regulators.

Discretion of the regulator
Regulating the performance of a private operator with a contract that may cover 10, 20, or 25 years requires that the regulator has room to manoeuvre and adjust the contract to unforeseen situations. In these circumstances, some level of regulatory discretion is desirable and necessary. On the other hand, a regulatory system that involves too

much discretion may deter private participation because it increases risk and arbitrary decision-making. To avoid this, it is necessary to ensure that: clear limits to discretion are specified in applicable laws and the contract; the criteria and processes to be employed by the regulator are established in law; and adequate provisions are in place for appeal against the regulator's decision. The definition of the discretion given to the regulator should provide assurances to operators and investors that regulatory discretion will be exercised in a way that protects their legitimate interests and does not subject them to undue political influence; to consumers that it protects their right to an adequate and safe service at a reasonable cost; and to elected officials that the regulatory agency will remain true to its mandate and accountable for its performance.

Regulator's independence
To be effective, the regulating agency must operate independently from both short-term political pressures and the regulated companies. If regulatory authority lies within the political control of government, there is always the danger that prices, service standards, and investment priorities will be manipulated to serve short-term electoral interests. With a more independent regulator, there is a greater chance that the sector can be managed to meet long-term service and efficiency goals that ultimately will lead to lower cost and better service. Achieving this goal is not easy but several safeguards could be employed, including: making regulatory appointments on the basis of professional, not political, criteria; appointing regulators for a fixed period; funding the regulatory body out of levies on utilities or consumers and not from government budgets; remunerating regulators competitively to attract and retain competent staff; and barring regulators from political activity and from having financial interests in water- and sanitation-related business. Several strategies could be used to reduce the risk of capture of the regulator by private regulated firms or political interests and to make best use of generally scarce regulatory skills: establish a multi-sectoral regulatory commission, for example one that deals with other related infrastructure such as electricity or telecommunications; contract-out some elements of regulation (such as financial auditing and monitoring service standards and asset conditions) to reputable competent private sector firms; use an existing regulatory body with a reputation for independence and honesty.

Accountability of the regulator

Although regulating agencies ought to have a high degree of independence from the political environment, there is still a need to ensure accountability. Ways to do that include: specifying the regulator's duty clearly in law; adopting transparent decision-making processes; requiring regulators to publish decisions and the reasons for those decisions; making decisions subject to review by the courts or some other independent forum; and requiring regulatory agencies to present annual reports on their activities and to be subject to independent audits.

Finding the appropriate regulatory system

In dealing with these issues, governments throughout the world have come up with greatly different solutions. Service and management arrangements are generally governed by detailed contractual arrangements managed directly by the utility or local government. Leases and concessions are usually subject to higher-level government oversight. In France, for example, where leases and concessions are largely a local matter, the national government has recently introduced national laws and regulations that govern important aspects of how these contracts are acquired by publishing model contracts and requiring competition. In Conakry in Guinea, the Société Nationale des Eaux, an autonomous state-owned national water authority responsible for water sector planning and investments, regulates the lease contract, but ultimate authority over tariff setting remains with the government. In Buenos Aires, an independent state agency representing local, provincial, and federal governments and funded from water and wastewater tariffs provides tight regulation, but its decisions can be repealed by government. In the BOO water treatment plant scheme in Sydney in Australia, the newly formed Sydney Water Corporation is overseen by the Pricing Tribunal of New South Wales, an independent authority that regulates retail water rates. In the United States, many states have utility commissions that regulate prices for an array of public utilities, including water. The Office of Water Services, the national regulatory body in England and Wales, is a national independent agency created specifically for the purpose of overseeing the privatized water and wastewater companies. In Chile, a national regulatory commission establishes price guidelines based on yardstick comparisons for all water companies in the country.

The role of the private sector

Governments interested in working with the private sector must realize that regulation is a critical part of any private sector arrangement. Basic decisions about the regulatory framework need to be made early. Regulatory capacity can determine which private sector option is most appropriate in a given situation. The regulatory system chosen can affect the willingness of the private sector to participate and the cost of its participation. There is no one right way to mix contracting and regulation or to define the most advantageous regulatory set-up. Options have advantages and disadvantages, and what works best will vary across countries and within cities in countries.

In developing a regulatory system, governments need to keep several broad principles in mind. First, the very purpose of regulation is to ensure that the interests of the consumer are protected. Secondly, the choice depends on the type of private sector intervention. A concession contract requires immensely more regulation and capacity to regulate than a management contract, where a simple contract may suffice. Thirdly, any choice must be realistic and compatible with the country's legal framework and its human resource capacity; a balance must be struck between the ideal and the achievable. Unless there is sufficient capacity, a concession may not be the way to go. Fourthly, regulation should not be too restrictive or controlling. Overly restrictive regulation could deter private companies from entering into private sector arrangements or limit their ability to introduce innovative and efficient operating practices. Any regulation that seeks to control in detail how the private contractor runs its business risks defeating one of the central purposes of private sector participation – improving the efficiency of service delivery by unleashing the know-how and creativity of the private sector. The very reason for choosing the private sector is that the private sector can operate in a less restrictive business environment compared with public entities. For example, if private participation is motivated by the desire to insulate the sector better from direct political intervention and to reduce public subsidies, giving regulatory authority to agencies dominated by short-term political interests would be counter-productive. If the private operator is subject to the same restrictive practices, standards, and norms that hobble public companies, even the best private operator cannot succeed. Without appropriate tariffs, for example, the private sector operator cannot raise the financial resources to meet its obligations. Governments must realize that bringing in the private sector cannot compensate for misguided and restrictive policies.

Preparing for a private sector venture

Governments interested in seeking a partnership with the private sector will have to choose from the many options available. These options can be implemented on different scales, with different combinations of functional responsibilities, and with different forms of regulation. The quality of the process of designing and implementing a selected option can determine whether or not the option will succeed. The challenge is twofold: (a) to define and develop the best possible arrangement appropriate for local needs and conditions; and (b) to find a suitable private sector partner for this arrangement and obtain the best possible offer from that partner. To reduce the risk of failure and get the most appropriate deal, the process is suggested to proceed in stages: pre-contract analysis; narrowing the choice to one or a range of feasible options; making sure that the desired option is of interest to the private sector; and acquiring the private partner and contract start-up.

Pre-contract analysis

Once a government has determined that private sector participation appears financially and politically feasible, a careful assessment should be undertaken before moving on to the next stage of choosing an option. Five main questions need to be answered:
1. What is the state of the existing utility?
2. How compatible is the existing regulatory regime with private sector participation?
3. How committed – or opposed – to private sector participation are key stakeholders?
4. Is the deal financially feasible for the private sector?
5. What are the main risks that need to be allocated or mitigated to ensure that private sector participation can succeed?

Assessment of the utility
The purpose of this analysis is to assess the current performance of the utility and the quality of information available about its management and operation and to identify those conditions that might make it attractive for a potential private sector partner. The government will need to gather – or identify as unavailable – information on matters such as: the utility's present and projected service area; the

current characteristics of the service (water quality, pressure, supply security, sewer flooding, or metering); a basic inventory of the assets and their condition and serviceability; human resources (numbers, skills, wage rates, conditions of service, pension arrangements); financial performance and tariffs (level and structure, subsidy arrangements, collection performance, disconnection policy); consumer preferences, affordability, and willingness to pay. The data collected in this evaluation will provide valuable information on the nature of the investment and related costs required to improve services, on potential efficiency gains that could achieved, on future desired performance standards, and on asset rehabilitation needs. A clear understanding of the utility's present performance is an essential condition for deciding on what form of private sector partnership would be desirable and feasible. It will also help to identify areas where data are lacking or may be inaccurate and lead to the conclusion that improving the data base may be necessary.

Regulatory/institutional analysis
The possible nature and effectiveness of a private sector arrangement depend on the regulatory mechanisms that will govern the private sector participation. Because any decisions about the private sector option, industry structure, and regulatory framework are closely linked, consideration of regulatory matters must be faced early on. Entering a private sector arrangement without an appropriate and effective regulatory structure can lead to costly mistakes and an acrimonious process to rectify matters later on. The objective of the analysis of regulatory issues is to: (a) identify elements in the existing broad framework of laws, constitutional requirements, and regulatory activities that could impede private sector participation or affect the viability of a specific option; (b) consider the potential for restructuring the present regulatory regime to open up the spectrum of options; (c) develop the sector-specific regulations that would govern the relationship between the public sector and the private partner; (d) help specify the powers that will remain in the public sector, identify who will exercise these powers and at what level of government, and create new regulatory arrangements as needed; and (e) decide which elements of regulation should be incorporated into the private sector contracts, how much the contracts limit the discretion of any public sector regulators, and what safeguards the contracts should contain against regulatory and political risk.

In assessing how the broad regulatory framework will affect the choice and design of a private sector arrangement and the attractiveness of that arrangement to the private sector, governments need to consider a wide range of laws and regulations related to: the constitutional and legislative division of responsibility for public services among national, regional, and local governments; inter-jurisdictional arrangements, if service responsibilities are decentralized and the system covers several jurisdictions; laws that govern private sector intervention in the provision of public services; water resource management and environment; labour; taxes; procurement of goods and services; currency control; public health and safety; and social policy.

Some elements of the existing framework cannot be changed or may take time to change. That may rule out a preferred option for private sector participation. If it does, it is best to recognize this early – and to adopt a stepwise approach to private sector participation that allows time to improve the general legal and regulatory framework. For example, if collection performance or requirements for providing subsidized services pose unacceptable revenue risks for the private partner, the best alternative might be to choose a private sector arrangement that reduces commercial risk by adopting a fee-based management contract type of arrangement. Or it might be possible to incorporate explicit safeguards into the contract – such as provisions allowing for additional payment for unexpected investments, protection from environmental liability, specified compensation or price adjustments for changes in service standards, and minimum revenue guarantees.

Stakeholder analysis
A range of stakeholders have a legitimate interest in the water business. Such key stakeholders and their potential support or opposition must be identified early. Stakeholders might oppose concessions or divestiture, for example, but accept management contracts, which give the private sector a more limited role. Or stakeholders might oppose any arrangement that has the private sector acting alone, but support joint ventures with the public sector. Generally these key stakeholders include: the national government (ministries with some jurisdiction over water-related matters, such as the ministries of health, environment, and urban and economic development); provincial and local governments that will act (or may act) as grantors of private sector contracts, regulators, partners, or financiers of the utility; regional or local planning departments, which coordinate land-use and infrastructure planning; other established regulatory

The role of the private sector

Table 7.3 **Potential stakeholder issues and policy responses**

Stakeholder group	Possible issues	Policy decisions required	Ways to get inputs
Employees	Staff redundancies; Changes in employment conditions	Redundancy packages and other arrangements encouraging staff to leave	Open and continuous consultations and negotiations with staff
Consumers	Consumers' preferences; Willingness to pay	System for planning extensions; Tariff methodology; Design of a subsidy scheme	Social Assessment, participation, public relations/consultation campaigns
Environmental interests	Major environmental consequences	Environmental standards to be applied; Liability for past pollution	Consultation with environmental groups
Existing government agencies	Major shifts in the allocation of responsibilities	Implementation of new regulatory system; Redefinition of responsibilities among government agencies	Intensive consultation
Other citizens	Resettlement	Resettlement policy	Direct consultation with affected groups

entities (such as water commissions, environmental agencies, and competition and fair trade commissions); political parties and individual politicians; labour unions; utility management and staff; consumer organizations; and advocacy non-governmental organizations concerned with some aspect of utility conduct and operations.

Once the key stakeholders have been identified, government will have to engage in a dialogue with all of them to gain support or to defuse opposition to the proposed private participation. To bring these stakeholders on board, government may have to consider a variety of conciliatory actions or safeguards (table 7.3), which may include protection for: (a) labour and management through redundancy packages, worker share allocations, minimum wages and working conditions, health and safety measures; (b) contractors or suppliers through regulatory rules to ensure competition in subcontracting and procurement; (c) customers through tariff adjustment rules, subsidy policies, complaint mechanisms; (d) public health and environmental protection through regulation of service standards, penalties for default. Although such safeguards can secure sufficient support to allow private sector participation to proceed and to ensure that it benefits users, they all involve costs that need to be carefully considered.

Among the most important stakeholders are utility staff and unions. Often, improving the efficiency of management and operations and introducing better technologies and know-how can be achieved only by restructuring personnel policies and reducing a costly and bloated workforce. Unless the staff's fear of layoffs can be managed, government's initiative is probably not possible.

The example of Buenos Aires in Argentina shows how government could deal with labour issues. The public water company in Buenos Aires, like many around the world, was heavily overstaffed. It had 7,600 employees, or about 8 employees for every 1,000 connections – around twice as many as needed to operate efficiently. Under a preferred plan, 1,600 employees accepted voluntary early retirement on severance packages financed by the government at a cost of around US$40 million. Soon after taking over the operation, the concessionaire offered another voluntary early retirement programme. This offer was accepted by 2,000 employees, at a cost to the concessionaire of around US$50 million. Within six months of the start of the concession, the number of employees had been reduced by almost half. The cost of achieving this reduction was considerable, but it was viewed as an investment necessary to achieve the efficiency targets sought by the concessionaire. Today, many of the former employees are among the 8,000 or so small contractors now providing services to the Buenos Aires company.

Analysis of financial and tariff conditions
Perhaps the most critical issue in selecting a private participation option relates to financial concerns. Unless the private sector feels that the deal is financially feasible, it will not enter. Typically, the following questions must be answered: If the private sector partner is expected to invest in rehabilitating the system or expanding coverage, how will that affect the tariff? Will the current tariff cover costs after

181

allowing for expected efficiency gains? If the projected tariff exceeds what some households are willing to pay, will the government provide subsidies? If not, could investment programmes be reduced to match the financial capacity of the consumers? Unless these issues are analysed seriously and realistically early on, much time and resources may be spent on options that turn out not to be financially feasible or not affordable. To find answers to these issues requires detailed financial work, including: assessing the financial status of the water and sanitation utility, and testing the financial and tariff implications of hoped-for service expansions and efficiency. Such financial analysis will narrow the options to arrangements that are feasible and sustainable.

Key parameters to be considered in this analysis include: (1) the utility's current operating and maintenance cost for water supply, treatment, and distribution, and for sewage collection and treatment; (2) current tariff levels and structure and collection efficiency; (3) current and projected water consumption; (4) the cost of capital improvements to the water supply, distribution, and sewerage systems, and annual expenditures necessary to achieve intended service levels; (5) the availability of funding for service improvements through grants, equity, and loans; (6) the additional annual operating costs due to system expansions, considering the efficiency gains that private operation might achieve.

Risk analysis
It is important for governments to recognize risks and to consider how they might best be allocated between the public and private sectors. Early thinking about the risks associated with private sector participation can save time later on and help ensure that the resulting private sector arrangement comes close to what was originally intended. Different risks are associated with different options, but they fall into two general categories. For fee-based service/management-type contracts, the most significant risk is that the performance of the private contractor may fall short of expectations. Arrangements need to be in place to monitor the contractor's performance, and to ensure that water quality and other standards required from the operator can be enforced. If adequate staff are not available to monitor performance, the government might consider contracting with a third party – an audit firm, for example. For lease/concession/BOT-type arrangements that require the private operator/investor to depend on

tariff revenues for financing operating costs and investments, incentives need to be in place for the contractor to improve the efficiency of operations and investments. Under a natural monopoly situation, there is always the concern and risk that the contractor will reap windfall profits by charging excessive tariffs or reducing service quality. These risks need to be managed carefully through the design of appropriate and effective monitoring and regulatory systems and arrangements.

The special problem of smaller cities
Private sector participation is easiest and most attractive in larger metropolitan areas, say cities with populations of at least half a million. Yet smaller municipalities have just as much need for better water and sanitation services and can also benefit from private participation. But their financial, economic, institutional, and technical conditions present difficult problems. A private contractor will often find it harder to make sufficient returns on its investments in small networks unless the operator can benefit from economies of scale by operating several smaller systems located close to each other. Also the generally lower average household income in smaller towns makes it more difficult for families to pay tariffs that cover costs and yield a reasonable return. Limited administrative skills and institutional capacity in many smaller municipalities limit their ability to undertake the effort needed to design, implement, and supervise private sector arrangements. Local officials and their staff will need assistance from higher-level government agencies in preparing for private sector entry. There are several ways to tackle these problems. The most effective is for several smaller towns to form an association and a single administrative entity, which provide the economies necessary to undertake private participation efficiently and competently. The national government can help smaller cities by supplying advisory services, financial models, and contractual documents.

Choosing among the options

Once a government has worked through the preliminary analysis it may begin to choose among options available. To narrow down the options, the following key questions must be answered:
- *What problems are to be resolved?* If the purpose of bringing in the private sector is primarily to improve operational efficiency and

The role of the private sector

management and administration, a management/lease-type contract would be most appropriate. If, in addition, investments are sought to increase service coverage and improve service quality, a concession may be the preferred solution. If there is need for a new plant, a BOT may be the best way of tapping private sector know-how and financial resources.
- *What are the implications for tariffs?* Do current tariffs cover costs? Can the private sector reasonably be expected to boost efficiency enough to meet the proposed service objectives without increasing tariffs? If not, will consumers be willing to pay higher tariffs? If not, will grant finance or subsidies to needy households be available?
- *Does the envisioned regulatory framework provide sufficient support for the private sector to take on commercial risk and at the same time protect consumers from abuse of monopoly power?* If not, can the necessary changes be made easily? If not, can parts of the regulatory function be simplified or contracted out in the short term? Can the risk of political interference be minimized?
- *Do key stakeholders support or at least not oppose a specific private sector option?* Can processes and policies be put in place to meet stakeholder concerns?
- *Is information about the utility's assets good enough to serve as a base for a long-term contract?* If not, can better information be produced rapidly or would it be better to take a staged approach: start with a management contract to assemble better information in preparation for a concession later on?

The answers to these questions will point governments to different choices or a range of choices for private sector participation. Table 7.4 depicts the relationship between private sector options and objectives. For example, a government seeking improvements in operating efficiency and responsiveness to consumers will prefer a management contract with performance incentives or a lease to either a service contract or a concession. A government seeking greater efficiency and new investment will prefer a concession or divestiture – or, for investment in bulk services, a BOT.

Making sure that the desired option is of interest to the private sector

A government's preferred option may not be attractive to the private sector. Where regulatory capacity is weak, political commitment low, the regulatory environments uncertain, and information poor, a con-

Table 7.4 **Private sector options and objectives**

Option	Objective				
	Specific technical expertise	Improving managerial capacity	Improving operating efficiency	Improving investment efficiency	Strengthening management autonomy
Service contract	Yes	No	Some	No	No
Management contract	Yes	Some	Yes	No	Some
Lease	Yes	Yes	Yes	No	Yes
BOT	Yes	Yes	Yes	Yes	Yes
Concession	Yes	Yes	Yes	Yes	Yes
Divestiture	Yes	Yes	Yes	Yes	Yes

cession or a BOT would be difficult to implement. Table 7.5 summarizes the enabling conditions that a private sector operator/investor would need to see fulfilled to agree to a given arrangement at a reasonable cost.

For a private sector option to be feasible, sustainable, and best suited to mobilize private sector skills and resources, it has to make sense technically, financially and politically. A technically sound arrangement is well targeted to the problem to be solved and is compatible with the existing legal framework – or includes supporting changes in that framework. A financially sound proposal is one that can be financed at a tariff that consumers are willing to pay – or with the aid of a fiscally responsible and politically viable government subsidy scheme. A politically sound proposal enjoys political support, both within government and among interested stakeholders. Experience around the world demonstrates that political commitment is absolutely crucial. To succeed, a political champion needs to provide support throughout the process of bringing in a private sector operator and/or financier in order to address sufficiently the concerns of key stakeholders. Political commitment is also essential to attract private sector interest. Potential private sector partners – and their financiers – will be looking for signs that the present government is willing not only to sign a contract but also to put in place regulatory arrangements that will protect their legitimate future interests. Unless there is a sense of political commitment, the private sector will provide less attractive bids or decide not to get involved.

Table 7.5 **Private sector options and enabling conditions**

Option	Stakeholder support and political commitment	Cost-covering tariffs	Requirement Good system information	Developed regulatory framework	Good financial rating
Service contract	Low	Preferred, but not necessary	Possible to proceed with limited information	Monitoring capacity for contract needed	Not necessary
Management contract with fixed fee	Low to moderate	Preferred, but not necessary	Possible to proceed with limited information	Monitoring capacity for contract needed	Not necessary
Management contract with performance incentives	Low to moderate	Preferred, but not necessary	Sufficient information required to define incentives	Moderate monitoring capacity needed	Not necessary
Lease	Moderate to high	Necessary	Good system information essential	Strong capacity for regulation and coordination essential	Not necessary
BOT	Moderate to high	Preferred, but not necessary	Good system information essential	Strong capacity for regulation and coordination essential	Better rating will reduce costs
Concession	High	Necessary	Good system information required	Strong regulatory capacity essential	Better rating will reduce costs
Divestiture	High	Necessary	Very good system information essential	Strong regulatory capacity absolutely vital	Better rating will reduce costs

Finding and contracting a suitable partner

Once a decision has been made on a feasible and appropriate option, the next task is to find and contract with a suitable private sector partner. Generally, the most effective way to do this is to require prospective partners to compete with one another to win the contract. The extent to which competition for a contract can be achieved – and the extent to which this competition translates into the best possible outcomes for consumers – depend on how well bidding is organized and managed. Several discrete steps are involved: the preparation of the bid documents, which communicate to prospective bidders the nature of the contract and the information on which to base their bids; organization of the bidding process and the way bids are evaluated and the contract signed; and selecting the private sector firms to submit bids.

Bidding documents and information for bidders
Based on the information prepared and decisions made during the pre-contract analysis, documents need to be prepared that clearly convey the information on which prospective operators and investors are asked to submit a proposal. The nature of the information to be provided depends of course on the option chosen. For a fee-based service or management contract the information requirements are relatively simple. Such a contract places little risk on the contractor. It could be bid much like a technical assistance contract, with the contractor's experience, proposed work programme, staffing, and cost as the basis for bid evaluation.

Long-term contracts, leases, concessions, and BOTs are usually awarded to the bidder proposing the lowest future tariff levels. For these kinds of contract, the operator/investor is running considerable risk when agreeing to a tariff arrangement that is to last for years. This risk can be reduced by having good information available that allows contractors to prepare their offers with good knowledge of what is to be expected. The more risk is involved, the higher is the need for good and reliable information. The better the information, the higher is the chance of concluding a contract that is realistic and responsive to both the government's and the public sector's requirements. The higher the risk is perceived by the contractor, the higher will be the price. If the risk is too high, the private sector may not be willing to get involved at all. For longer-term contracts, the bidding package should include the following information: (a) technical information: a detailed description and assessment of the existing facilities and asset analysis, improvements to be obtained under the contract in terms of service improvements and performance standards, and an estimate of capital expenditures; (b) legal and regulatory information: a clear definition of the legal and regulatory system under which the contract will be done and the provisions that will be covered under the contracts; (c) economic and financial information: demand projections and willingness-to-pay analysis, policy on tariffs and tariff structure, the proposed capital structure of the deal – in short, all the information that prospective contractors/investors would need to perform their own financial analysis and forecasts; (d) human resources information and the policy to be implemented by the contractor; (e) the rules and scoring mechanisms that will be used to evaluate bids and deal with complaints and appeals; and (f) a timetable for bidding, evaluation, and award.

Operators/investors bidding for long-term contracts that require a financial commitment will carry out their own analysis. The less confidence the private sector has in the veracity of the information provided by the public contracting entity, the more exhaustive and time consuming this analysis will be. For large contracts involving decisions on hundreds of millions of dollars, private partners may spend several million dollars in making their own assessment of all aspects of the future project to make sure that their offer is based on a realistic understanding of the task at hand and the risks involved. These costs can be reduced by having good and reliable information available to bidders and limiting the number of bidders invited.

Organizing the bidding

Experience suggests that the best way of finding a suitable partner at a reasonable cost is to require competition among prospective partners. There is a variety of possible contract bidding and award procedures, which can be grouped into three categories: competitive bidding; competitive negotiations; and direct negotiations. Each has its advantages and disadvantages.

A competitive bidding process involves a formal, public process for presenting proposals, evaluating them, and selecting a winner. Its main advantage is that it ensures transparency, provides a market mechanism for selecting the best proposal, and stimulates interest among a broad range of potential partners. It works best where outputs are standardized and all technical parameters can be clearly defined. To avoid misunderstandings and avoid underbidding, the quality of information must be high, an undertaking that is often difficult given uncertainties and the long time-frame – 20–30 years for a concession – involved. Also, direct competition limits the scope for the presentation of innovative proposals that deviate from the base requests made in the bidding documents. These issues do not mean that competitive bidding should be avoided, however, but particular attention should be paid to providing good-quality information to potential bidders and to the detailed design of the bidding process.

If competitive negotiations are used, the government invites proposals from selected bidders to meet specific service objectives. After review of the proposals based on their technical merit, the government negotiates contract terms and conditions with a small number (usually two or three) of selected bidders. This involves simultaneous negotiations with these bidders. Competitive negotiations are well suited to projects in which many technical variations are possible and the contracting entity wishes to explore different and creative proposals without being bound by the standard solutions that a competitive bidding process would require. They offer a richer means of considering other factors than price. Competitive negotiation has some risks, however. It is less transparent than competitive bidding and may raise concerns about corruption and favouritism. The government can reduce the risk of any impropriety by specifying publicly, and as clearly as possible, what the evaluation criteria will be, by standardizing the negotiation processes across bidders, and by keeping a detailed record of the process.

Direct negotiations occur most often where a project idea originates with a private sector sponsor rather than with the government. A developer or operator seeks to negotiate directly with a government or a public utility the terms and conditions for a management contract, BOT, or concession. Direct negotiations can be a good way of attracting innovative projects and securing private sector involvement in smaller cities and towns where the costs of entering competitive bidding contests may be high relative to the expected returns. But direct negotiations make it difficult to ensure transparency in the selection process and an efficient outcome. Without competition, it is much harder to assess the reasonableness and cost-effectiveness of a proposal. And direct negotiations, even more than competitive negotiations, can be associated with corrupt and improper behaviour of the public agencies carrying out the negotiations. Allegations of improprieties, whether they are well founded or not, can lead to resistance from key stakeholders and in the extreme to the eventual cancellation of the contract. If direct negotiations are contemplated, governments must take extra steps to ensure transparency and efficiency. For example, a government might establish an independent advisory panel to advise on whether direct negotiations are appropriate for a particular project. It may require all contracts to be reviewed by a national or regional regulatory entity that may use benchmark comparisons of construction costs or service tariffs to assess the efficiency and appropriateness of the negotiated deal.

In general, the more competitively and transparently the selection of the private partner is conducted, the greater is the likelihood that the best possible deal will be achieved and that the deal will be politically sustainable. For these reasons, most governments – and also multilateral agencies such as the World Bank – favour or require competitive bidding for private sector contracts. Many countries have laws that explicitly forbid direct negotiations. However, it should be recognized that there may be circumstances that make it difficult to achieve perfectly competitive bidding. If information is incomplete, for example, or there is a range of possible solutions to the service problems the government is trying to solve, the government may wish to enter into a dialogue with potential bidders to work out how best to specify the contract. This approach does not preclude competition, but it does reduce transparency and the chance that bidders will be able to bid on equal terms. Direct negotiation is clearly less preferable than competitive bidding, but in some situations it may be a feasible approach. For example, the costs of competitive bidding can be so high relative to the expected revenue stream from small contracts as to deter bidders. In these cases, governments ready to undertake direct negotiations would be well advised to employ special safeguards, processes, and auditing procedures to ensure that the best possible partner is found on the best possible terms, and that the resulting contract will stand up to political and technical scrutiny.

Pre-qualification of bidders
Pre-qualification is strongly recommended for all types of contract options as a way to ensure that potential bidders have the technical and financial capacity that the task demands and a track record in performing similar tasks. It is important to weed out firms that clearly do not have the capacity to take on the job before they prepare costly proposals; once they have entered the evaluation process, they may bring political connections to bear to win the job. Reducing the number of bidders also reduces the cost of preparing proposals,

which, as outlined above, may be several million dollars for large projects involving the provision of substantial amounts of private capital. Limiting the number of bidders will increase a private firm's incentive to participate in the bidding, because it increases each bidder's chance of winning. Faced with a dozen competitors, some of them with questionable credentials, most qualified firms may choose not to participate in the bidding.

Pre-qualification criteria generally include some combination of: financial capacity; relevant experience; and past record on similar ventures. The criteria for evaluating firms participating may be either qualitative or quantitative. Qualitative criteria allow greater flexibility and discretion, but they are also less transparent and more likely to produce complaints by bidders that fail to pre-qualify. Again a balance needs to be struck to ensure a fair and transparent process. In defining pre-qualification criteria for water and sanitation contracts, governments need to keep in mind that the number of private companies with substantial experience in providing water and sanitation services is relatively small. Few firms today meet stringent and ambitious pre-qualification criteria. To broaden the range of potential bidders, governments may consider firms other than water and wastewater operators. For example, a telecommunications company or a company with experience on the commercial side of electricity distribution might be able to handle the commercial side of a water business when paired with a company with engineering expertise in the sector.

Pre-bid contacts with bidders
In deciding what form a private sector arrangement should take, governments may want to know early on the opinion of the private sector. For example, a government might want the private sector to make large investments in new capacity and take all the commercial risks associated with them – only to find that the private sector judges the risk to be too high. Or a government might assume that local circumstances are so unattractive that the best it can hope for is a fixed-fee management contract – and unknowingly preclude initiatives by private companies that would be prepared to take more commercial risk. To avoid these misunderstandings it is generally a good idea to have informal discussions with bidders before finalizing the bidding documents. Bidder feedback on early drafts of the bidding documents or regulatory design can help identify changes that would make the transaction more attractive to private firms with no loss to the gov-

The role of the private sector

ernment or other stakeholders – and result in better, more affordable bids. The government must assure, however, that all prospective bidders receive the same information, to avoid complaints that some bidders were favoured over others.

Bid contents and evaluation
Central to the bidding process are decisions about what (pre-qualified) bidders should be asked to include in their bids and how these bids should be evaluated. Traditionally a two-stage bidding system is used, requiring bidders to submit a technical envelope and a financial envelope.

The technical envelope may have various purposes varying in complexity and transparency. In the simplest case, the technical envelope contains legal certification of the bidding consortium and a bid bond. Once these items have been confirmed, the financial envelope is opened and the contract is awarded to the best offer. In a second approach, the technical envelope serves some of the purposes of pre-qualification – if pre-qualification has not taken place earlier – and provides technical and financial information on the bidder. Some bidders may be disqualified once this information is assessed. The financial envelopes of the surviving bidders are then opened, and the contract is awarded to the best offer. These two approaches are relatively simple and transparent. They tend to work well when technical requirements can be specified. A third approach requires bidders to include a technical proposal that sets out the proposed business plan (including investment and financing plans) for meeting the service objectives. The plans are reviewed for consistency with the project specifications and requirements, and proposals either pass or fail. Again, the contract is awarded to the surviving bidder with the best financial bid. This approach was used for the Buenos Aires water concession. Under a fourth approach technical proposals are required as in the previous case but, rather than passing or failing, the proposals are scored. The financial proposals are also scored, and the contract is awarded on the basis of the weighted technical and financial scores. This approach was used to allocate freight rail concessions in Argentina. These latter two approaches are more complex and less transparent. They may be preferred when the technical criteria cannot be clearly specified in advance and when the government is looking to bidders to present their own ideas on how to achieve service objectives. The third approach might be chosen if the government has firm and clear ideas on the minimum technical requirements; the fourth if there is less clarity about requirements, and if different technical proposals may have different financial implications at different stages of the project's life. For these more complex approaches, the government should specify as clearly as possible and in advance the processes and rules that it will use for evaluating bids.

The financial envelopes provide information on the financial conditions under which the operator/investor offers its services and bids are awarded. They also vary in form and complexity, depending in part on the form of private sector participation. For management/service-type contracts, bids are awarded to the bidder that quotes the lowest service fee. For concessions or BOTs, the financial envelope contains the bidder's proposed future service tariffs for which it would be prepared to enter into a concession or the take-or-pay fee for bulk supply. This approach was used, for example, for the Buenos Aires water concession. Bids may also propose an up-front payment in combination with future concession fee payments. This approach is appropriate for concessions and leases and was used, for example, in the Argentine freight rail concessions. The bid is evaluated on the basis of a weighting of the up-front payment and future fees. In the case of privatization involving the sale of shares or the divestiture of assets, bids present a price of the shares or assets being sold.

Complaints and appeals

The more complex a bidding process, the greater the chance that competition will be perceived to be unfair or that losers will question the choice of winner. The first-best solution to such problems, of course, would be to make perfect information available to all the bidders, have truly unambiguous bidding rules (the lowest price or tariff wins), and preclude substantive negotiations after the bidding contest. For obvious reasons this is rarely possible. The next-best solution is to structure the process as clearly as possible, ensuring that everyone has access to the same information, that bidding and evaluation rules are as simple as possible and are clearly explained at the outset, and that there are clearly defined limits on post-bid negotiations. But no bidding process, no matter how carefully structured, can eliminate the potential for complaints and appeals. Consequently, the government should create a mechanism for handling complaints, specifying: who will be responsible for hearing and arbitrating complaints and appeals; on what basis complaints and appeals will be heard; how complaints and appeals should be formulated; whether a fee will have to be deposited for each complaint to discourage frivolous complaints; and what the deadlines are for the receipt of complaints and appeals and their resolution.

Managing the process

Creating a management unit

Competent management of the entire process from the pre-contract analysis through to the signing of the contract is essential for success. The government entity leading the effort should establish a unit responsible for the day-to-day management of the process. The skills of the people appointed to this unit will be critical. In deciding how to set up the unit, the government's objectives should be to: (a) ensure that the unit has sufficient autonomy, both managerial and financial, to carry out its task cost-effectively; (b) shield the unit's staff from political interference in their day-to-day tasks; and (c) give politicians and relevant government agencies confidence that the task is proceeding as directed and that any major policy issues are dealt with properly by putting in place reporting and accountability mechanisms.

Hiring competent independent advisers

Designing and implementing private sector participation in water and sanitation require substantial economic, financial, technical, and legal expertise and the coordination of that expertise. The process requires detailed work – first refining the option to be implemented and the legal and regulatory measures needed to support it, then preparing many complex documents, such as the regulatory framework law, the bidding documents, and the draft contracts. Governments usually lack the range of expertise within the civil service to carry out all of these tasks. Even where earlier privatization projects may have helped to build up a body of some skilled staff, government would be well advised to engage external advisers. Managing these advisers then becomes a primary task of the government unit.

The type of advisers generally needed include: *economic and regulatory consultants*, to advise on how the deal might be structured, competition might be promoted, and tariffs might be structured and adjusted, and on what regulatory and monitoring mechanisms are needed; *legal consultants*, to deal with legislation and regulatory issues and prepare bidding documents and draft contracts; *technical consultants and engineers*, to undertake the technical assessment and prepare technical specifications and requirements for the contract; *environmental consultants*, to prepare environmental studies; *investment bankers and financial consultants*, to prepare financial projections and determine the financial feasibility and structure of the contract. Hiring competent, independent advisers who can assist the government throughout the process is probably the best investment the government can make. Quality advice is expensive, but signing a poorly structured or inappropriate deal will be much more costly later on.

Time requirements

The time required to complete all the elements on this critical path will vary among countries and according to the type of private sector option being pursued. Countries with legal and regulatory frameworks supportive of private sector participation in water and sanitation and with good-quality information on the system may be able to proceed relatively rapidly. Management contracts should take less time to prepare and implement than concessions: given strong political commitment, a management contract could be designed and implemented in 8–10 months, whereas a concession could easily require 18–24 months. The Buenos Aires concession, for example, took two years to prepare, and the Manila concessions were completed in around 18 months. As a country undertakes more private sector contracts, it may be able to shorten the preparation time. But there is a lower limit, determined by the need to develop an arrangement well tailored to local circumstances and by the time required by potential bidders to develop considered offers.

Managing the contract

Contract closure marks the beginning of an ongoing partnership between the public and private sectors. The quality of the contract and the quality of the contractual partner are very important to the success of this partnership, but so too are the institutional arrangements put in place for maintaining and governing the partnership and for perpetuating competitive pressure on the private partner. Most public–private partnerships are long, and planning for their maintenance is critical. The poorer the quality of information at the start and the greater the doubts of both parties about their relationship, the more inevitable renegotiations of the original deal will be and the more important it is to introduce robust provisions for renegotiation and to supplement competition at the bidding stage with future competitive pressures.

Contract renegotiation

Most initial contracts are based on incomplete information. But even if a contract were bid on the basis of perfect present information, the future holds uncertainties that cannot be handled by contract. So careful provision must be made to deal with unexpected events over the life of a contract. These provisions can turn out to be even more important to success than the initial terms of the contract. The issues that should be covered by contractual and institutional provisions for renegotiation and adjustment of contracts include four general essential elements: the conditions under which adjustments to the contract terms can be made; guidelines determining when and under what conditions contract *renegotiation* must occur rather than price or service adjustments by agreement or by regulatory discretion; the process by which renegotiation must be initiated and conducted; and the definition of clear arbitration provisions.

Maintaining competitive pressure

In the water and sanitation sector, where monopoly power is inevitable, one important function of the regulatory system is to attempt to ensure that private companies operate as efficiently as they would have to in competitive markets. Some competitive pressure is introduced when companies compete to win a private sector contract, but it is short lived. Regulators can exert longer-lived competitive pressure by: (a) allowing direct competition, say at the boundary of a concessionaire's area or for specific new services within its area, and by ensuring that major new capacity expansions are not simply negotiated with the incumbent but are bid for; (b) using yardstick or comparative competition to monitor and assess the performance of the private sector operator; and (c) choosing a form of price control that explicitly requires the company to make efficiency gains.

Preventing undue outside interference

Once a contract has been awarded to a private company, it is the job of that company to run the business in accordance with the agreed contractual and regulatory conditions. This may seem an obvious point. But experience suggests that great care is needed to ensure that regulators do not become involved in the day-to-day management of the utility. Regulatory tasks – and regulatory staff – need to be focused on desirable *outcomes*, not on how to achieve these outcomes. For example, it is the regulator's task to specify a standard for drinking water quality and to establish a system for monitoring performance against this standard. It is the company's task to decide what technical measures and operating practices are needed to meet the standard. When a government specifies the regulator's duties and decides on the appropriate staffing and skill mix for the regulatory agency, it must have a clear understanding of the dividing line between regulation and operational management.

Conclusions

One of the great challenges for the twenty-first century is to provide present and future populations with adequate and safe water and sanitation. In addition, the protection of the world's water resources from unsustainable exploitation and the safeguarding of the aquatic environment require prime attention. The financial resources required to provide facilities for the many who do not have access to these services today and for future generations to come are enormous. Financial resources alone are not sufficient. Fundamental improvements in human capacity to manage and operate these services better are also needed. To this effect, past and present restrictive sector development policies based on the public sector have proven to be ineffective. The greatest need for reform is in middle- and low-income countries where both financial resources and capacity are more limited than elsewhere.

The private sector as provider of services and materials, and also as operator and financier of water and wastewater infrastructure, will have to play an important role in the future development of the sector. Involvement of the private sector is not new. In some countries it has been working for more than a hundred years. Recently, countries and cities throughout the world, including those in low- and middle-income countries, have increasingly looked to the private sector for expertise and financial resources. Yet many decision-makers are reluctant to seek the assistance of the private sector. In great measure this reluctance has its cause in misunderstandings and insufficient understanding of how the private and public sectors can work together for mutual benefit. The following conclusions summarize the salient points to consider in making decisions regarding the involvement of the private sector in municipal water and wastewater.

- Many examples exist throughout the world that demonstrate clearly that private sector participation can make significant contributions to improving the efficiency of utility management and operations and providing capital for investments; among many examples, the Buenos Aires concession stands out.
- The private sector can participate in many different ways, ranging from simple service contracts to leases and concessions to full divestiture; in these arrangements the division of responsibility and risk between the public and the private sector varies widely. Need and financial and legal conditions determine which of the many options are feasible and sustainable in a particular setting; careful analysis is needed to choose an appropriate option.
- Because of the natural monopoly features inherent in the provision of water and wastewater services, the private and public sectors must form partnerships. Enabling contractual and regulatory structures and relationships are essential to ensure that: (a) the private sector discharges its responsibilities efficiently and receives a fair return on its investment; and (b) the public sector monitors the performance of the private sector to ensure that the private partner discharges its responsibilities as agreed. Appropriate regulatory/contractual arrangements are of key importance in structuring the relationship between the public and private sector.
- The choice between the various private participation options depends on the problems to be resolved, but also on the financial capacity, stakeholder opinion, and the legal/regulatory environment in a country. The presence of a political champion is essential.
- The greater the involvement of the private sector in terms of taking on performance responsibilities and commercial risk, the higher the need for an enabling environment that provides comfort to the private sector.
- The best way to acquire a "good deal" is through competition among prospective qualified private providers of services; competition is the best way to ensure the lowest-cost deal and to dispel any notion of impropriety associated with a sole source negotiated deal.
- Introducing the private sector and choosing the most advantageous option are complex tasks, particularly in countries where no previous experience exists. Careful analysis is need to determine feasible objectives, deal with technical, financial, and regulatory issues, and

prepare for and implement the bidding and contracting process. Competent and independent advice from consultants is essential to ensure a successful outcome, i.e. a lasting relationship beneficial to all parties.

Acknowledgements

The views expressed in this paper reflect those of the author and not necessarily those of the World Bank.

The author also wishes to acknowledge that, in preparing this paper, extensive use was made of information contained in the June 1997 draft of the "Toolkits for Private Sector Participation in Water and Sanitation," prepared under the leadership of the World Bank's Transport, Water and Urban Development Department. The toolkits were subsequently published by the World Bank.

8

Emergency water supply and disaster vulnerability

Charles Scawthorn

Introduction

While the twenty-first century is anticipated to be characterized by major population and mega-city growth,[1] with many technological and social advances, one thing that will not change is the threat of great natural disasters, such as earthquake, wind, drought, and flood. Urban water supply systems are vitally required for day-to-day existence, commercial and industrial operations, and emergency firefighting. With major urban growth, mega-city water supply will become more vulnerable, owing to water being conveyed over great distances by only one or a few major canals or pipelines. Tokyo, Los Angeles, Mexico City, Beijing, and other mega-cities may thus be faced with loss of drinking water for extended periods as a result of natural catastrophes, which in some cases can be accompanied by great fires or other additional demands for water.

Loss of water supply during disasters has led to severe public health problems and occasional conflagrations and catastrophic losses (Scawthorn and O'Rourke, 1989; Scawthorn, 1996a), which might have been avoided given an assured and adequate, or *reliable*, water supply. Reduction of the potential for loss of urban water supply due to natural disasters is a difficult task, but can be accomplished via a judicious combination of: planning; development of disaster-resistant structures and other features in the water supply system; and development of alternative potable water supplies, or delivery systems balanced for the specific situation. The purpose of this paper is to describe recent developments that provide insights into the appropriateness of each of these technologies.

Development of a reliable water supply

In order to find the appropriate balance of technologies, several steps are required. These steps in effect constitute the development of a reliable water supply system, and are shown in figure 8.1. They consist of:
- Defining **performance criteria** for the water supply system – that is, *how much* water, of *what quality*, and with *what reliability* of delivery is required?
- Defining the range of options available for the collection and conveyance of the water – that is, what are the **alternative design** measures and schemes?
- **Analysis** of the alternatives to determine the probability, or reliability, of the system attaining the performance criteria, under these various alternatives.
- **Cost benefit** analysis, to find the optimum combination of alternatives attaining the performance criteria.

Of these major steps, perhaps the most elusive has been the definition of performance criteria – that is, *how much is enough* (in this case, water)? In general, the performance goals for water systems are to provide sufficient supply for: (a) potable supply (desirable within *hours*, although can be tolerated for *weeks*), (b) timely recovery of industry (*days to weeks*), and (c) occasionally, immediate fire suppression (*delivery required within minutes*). The remainder of this chapter seeks to inform the reader about the present status of this problem, by reviewing the performance of water supply systems in several recent disasters, providing an overview of water supply reliability methods for the lay reader, describing several case studies where water reliability or other technologies have been applied to reduce potential disaster vulnerability and assure emergency water supply, and, lastly, presenting a proposal regarding an international solution to this problem.

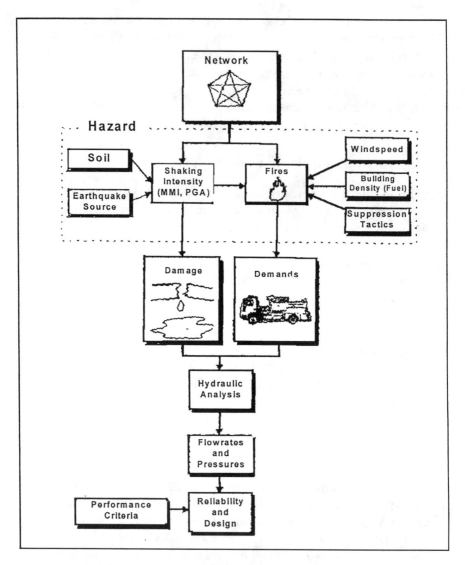

Fig. 8.1 **Generalized methodology for water system reliability analysis (Source: Scawthorn, 1996a, after Scawthorn et al., 1995)**

Performance of water supply systems in recent disasters

Water systems have historically been vulnerable to natural hazards; the great disasters of San Francisco in 1906 and Tokyo in 1923 are two of many examples. This vulnerability however is not confined to the historical past – modern water supply systems have performed unevenly in recent incidents. In order to highlight the vulnerability of our modern and future cities and some of the key deficiencies, I briefly review here the performance of water supply systems in selected earthquakes, fires, flood, and refugee situations, all during the past decade.

Earthquake

California, 1989 and 1994
The 1989 earthquake in Loma Prieta, California, which measured M7.1,[2] was a significant event that caused a partial collapse of the San Francisco–Oakland Bay Bridge (one of the world's largest bridges), the loss of 1,000 ft (300 m) of airport runway owing to liquefaction at the Oakland International Airport, and the collapse of modern freeways and buildings throughout the San Francisco Bay Area. A number of water systems were affected:
- the San Francisco Municipal Water Supply System (MWSS) sustained about 150 breaks, many in the Marina (Scawthorn and Blackburn, 1990), where service was disrupted for weeks;
- a number of breaks were sustained by the East Bay Municipal Utility District (EBMUD) (Eidinger et al., 1995);
- the special aseismic San Francisco Auxiliary Water Supply System (AWSS) sustained very few breaks, but performance was still impaired for several hours owing to operational errors; although this performance conflicted with its specialized aseismic function, the system responded adequately to a potential conflagration within one hour of the earthquake, and service was restored rapidly enough to be available had a major conflagration developed (Scawthorn et al., 1990).

The 1994 Northridge M6.7 earthquake was also a significant event, affecting the greater Los Angeles region. As with the 1989 event, several major freeways and buildings collapsed, but significant earthquake strengthening programmes had been in place since before the 1989 event, and their cumulative impacts reduced the potential damage. Nevertheless, the total monetary damage for this event has been estimated at US$40 billion, making it the largest natural disaster in modern US history in financial terms. Water service was lost in a major portion of the San Fernando Valley region of Los Angeles as a result of about 2,000 breaks, and LAFD was forced to resort to drafting from backyard swimming pools to combat an estimated 109 ignitions. Although there were many fires, no conflagration developed, owing in part to favourable conditions but also in part to the fires being suppressed early given the modest water supply from these swimming pools. Overall, service was restored within several days (Eguchi and Chung, 1995).

Japan, 1993 and 1995
Relatively small communities were affected by the 1993 Okushiri M7.4 earthquake event in northern Japan, but water supply aspects were notable because the Aonai village system supply was lost at a river crossing because of pipe breakage and, when a conflagration developed, fire-fighters were prevented by tsunami debris from drafting from the ocean, although it was only a few hundred metres distant (Yanev and Scawthorn, 1993).

The 1995 Hanshin M6.9 earthquake is probably the most significant earthquake to have affected a modern industrialized society in recent times. Many modern buildings, bridges, and freeways collapsed, as well as thousands of homes, resulting in over 6,000 fatalities. The total damage for this event has been estimated at US$100 billion, making it the largest natural disaster in modern history in financial terms. Water service was lost in virtually the entire urbanized area of Kobe as a result of about 2,000 breaks. The fire department was confronted with about 110 fires very soon, and was unable to relay adequate water, resulting in several conflagrations. Potable supply was significantly impaired: service was restored to 50 per cent of customers in 10 days, but 20 per cent were still without supply one month later (Ballantyne, 1995).

Fire

The 1991 Oakland Hills and the 1993 Southern California fires are examples of conflagrations requiring large amounts of water for an adequate response. When the Oakland Hills area was first developed, heavy vegetation and numerous trees were planted in conjunction with extensive home-building. These brush zones contributed to the intensity of the 1991 conflagration. The hilly area required special pump stations and hill-top tanks for water supply. Because the pumps were not of large capacity, the tanks were soon exhausted and there was no water supply available for suppression. In Southern California, the Santa Ana winds combine with extremely flammable brush in the mountains and the wildland–urban interface to create explosive fire conditions, which have resulted in major conflagrations at regular intervals in Southern California's history.

Flood

On Sunday, 11 July 1993, the Des Moines Water Works treatment plant in Iowa was submerged by floodwaters from the Raccoon River, leaving Des Moines without drinking water and fire protection. The plant treats water for 250,000 people. The floodwaters receded within two days, the plant site was dewatered, and the first finished water pump motor was pulled for reconditioning. The treatment plant was back in service by 16 July, and water was being pumped into the system the following day, seven days after the submergence. The distribution system was filled and pressurized within 12 days, and water quality was restored for drinking 19 days after the treatment plant was flooded. Final restrictions were removed 29 days after the initial flooding. There was a total of US$12 million damage, US$10 million of which was covered by insurance. The Federal Emergency Management Agency reimbursed most of the balance. There was an additional loss of US$2 million in lost revenues.

In April 1997, the Grand Forks metropolitan area in North Dakota (total population 100,000) was devastated by a major flood requiring the evacuation of 60,000 persons. The Red River of the North, the source of the floods, was also the sole water source for the community, and mechanical and electrical equipment in riverside water treatment plants was severely damaged. However, as with Des Moines, the water service was restored relatively rapidly, within several weeks. Total damage for this event is estimated in excess of US$1 billion.

Refugee camps

Starting on about 13 July 1994, there was a sudden influx into Zaire of 1.5 to 2 million Rwandan refugees, owing to a civil war in the Rwanda area (Biswas and Tortajada Quiroz, 1995). Camps were hastily sited and, in some cases (like Kibumba), no water was available. This was despite the fact that the availability of clean drinking water is a priority requirement of UNHCR camp-siting policies (UNHCR, n.d.). On the basis of a field reconnaissance, Biswas and Tortajada Quiroz (1996) noted:

Because of the volcanic nature of the land around the camp, all drilling attempts by Oxfam and SIDA (Swedish International Development Agency) to find water have so far been unsuccessful... In terms of current health problems, the main one is diarrhoea. Dysentery is being reduced and occurrence of cholera has not been observed for some time. Malaria (Plasmodium falciparum) has been a chronic problem. All these diseases are water-related.

When the camps were set up in late July, cholera had been the cause of major loss of life, which is described in the section on "Applications" below.

These examples illustrate the following points:
- it is very difficult to avoid all disaster damage,
- general performance criteria can be achieved, even with considerable damage,
- when performance criteria are not achieved, catastrophic damage to water systems and loss of life can occur.

As outlined in the introduction, mitigation of this potential for catastrophic damage and loss of life requires the development of reliable water supply systems. Several steps are required for this, so the next section provides an overview for the lay reader of water supply reliability methods.

Overview of water supply reliability methods

Adequate water supplies can be assured (i.e. made reliable) via an optimum design programme involving: robust hardware, alternative flexible emergency response resources, redundant supplies and networks, and real-time control systems.

Although urban water supply systems are notable for the high degree of redundancy of their distribution networks, they are also, in a larger sense, significantly serial in nature. That is, many urban water supply systems are typically serially connected sub-systems, with many cities having only one major supply/watershed, connected by transmission along only one right-of-way, and with only one or a few treatment plants prior to connection to terminal reservoirs and the highly interconnected distribution system. Of course, each of these sub-systems (especially the distribution system) is more complex. For the overall serially connected system, or any of the sub-systems, two types of analysis can be performed:

(a) **Connectivity analyses** measure post-earthquake completeness, "connectedness," or "cut-ness" of links and nodes in a network. Such analyses ignore link, node, or system capacities and seek only to determine whether, or with what probability, a path remains operational between given sources and given destinations.
(b) **Serviceability analyses** seek an additional valuable item of information: if a path or paths connect selected nodes following an earthquake, what is the remaining, or residual, capacity between these nodes? The residual capacity is found mathematically by convolving link and node capacities with network "connectedness."

Thus, serviceability analyses seek to determine adequacy. For the simplest series system (assuming each element is independent),

$$\boxed{\text{START}} \text{---} \boxed{A} \text{---} \boxed{B} \text{-----} \boxed{n} \text{---} \boxed{\text{End}}$$

the probability that Start will be *connected* to End (denoted P_s = probability of survival) is simply:

$$P_s = P_s^A \cdot P_s^B \cdot \ldots = \prod P_s^n,$$

where P_s^A is the probability of survival of link A, etc. The probability that flow greater than x will be transmitted from Start to End, or the probability that the system will be *serviceable* for flow x, is similarly the product of the probabilities that each link A, B, ..., n will transmit flow greater than x or be serviceable to that level (note that this treats the links as independent). For parallel and especially for more complex networks, Monte Carlo techniques are typically employed, especially for serviceability analyses (see Scawthorn et al., 1993, for a review of the methods employed in these types of analysis).

Using this type of analysis, the probability of non-serviceability or failure, $P_F = 1 - P_s$, can be determined, and combined with the consequences of failure to determine the expected cost of failure. For example, analyses may show that a number of ignitions will occur as a result of an earthquake, and that these ignitions will grow into a conflagration of dimensions such that N buildings and property will be destroyed by the fire. If the analysis also determines that x flow of water is required to prevent these ignitions from growing into a conflagration, then the expected cost of failure given the event is $P_F \cdot N$, where P_F is the probability of failure to provide x flow of water. The expected cost of failure can be compared with the cost of stronger or additional pipe, or other mitigation measures, in a cost–benefit analysis to determine the optimum reliability-based design. Note that in this cost–benefit analysis, the avoided cost of failure is the "benefit" (which is derived by investing in the improvements of additional pipe, etc.).

Applications

The techniques outlined in the previous section have been employed in recent projects in an effort to assure reliable water supplies in several cities. Each case was unique, but all the projects began with identification of existing vulnerabilities and determination of the urban region's appropriate reliability requirements.

Contra Costa Water District

Contra Costa Water District (CCWD) serves approximately 400,000 people in the San Francisco Bay Area (Contra Costa County), drawing its water from the Sacramento Delta and transporting it along the 77 km Contra Costa Canal (open channel). The basic CCWD system consists of the canal conveying water to the Bollman Water Treatment Plant (BWTP), whence treated water enters a distribution network that includes a number of pressure zones with associated pump stations, tanks, etc. The majority of demand is in the western portion of the District, whereas the water is drawn at the extreme eastern edge. Average daily demand for the entire system is approximately 454,000 m³ (120 million gallons/day, mgd) (Avila and Chan, 1995). Because of population growth, the CCWD planned capital improvements of about US$160 million but, considering it is located in a seismic region, wished to take earthquake and other hazards into consideration in its capital improvements programme.

The first step in the reliability-based design for this system consisted of defining *stress events*, which included both regional stress events (earthquake and associated fires, wildland–urban interface fires, the spillage of hazardous materials, and regional power loss) and local stress events (industrial fire, windstorm, local power failure, explosion/transport accident, landslide). Preliminary analysis showed that fires would be serious stresses on the system reliability. Earthquake events considered in the analysis were an M7 Coast Range

Table 8.1 **CCWD system qualitative reliability summary: Fire following M6.5 Concord Fault earthquake**

Facility	Reliability
Existing raw water system	
Canal	Moderate
Raw water pump stations	High
Reservoir	Low
Existing treated water system	
BWTP	Low
Storage tanks	Moderate
Pump stations	High
Pipelines	Low

Sierra Block and a nearby M6.5 Concord Fault event. The M6.5 Concord event was found to result in likely damage to selected CCWD facilities and about 40 serious fires, with a total required fire flow of about 97,000 gallons/minute (gpm; 388,000 litres/minute). Methods used for the fire following earthquake (FFE) analysis have been presented previously (Scawthorn, 1987).

A serviceability/reliability analysis as defined above was performed for the system consisting of the canal, the BWTP, and the distribution network (including tanks, pump stations, etc.). The findings were that the probability of meeting FFE demands (i.e. reliability) varied by service zone, from very low (<10 per cent) to moderate (>80 per cent). Qualitative system component reliability is shown in table 8.1. System reliabilities for wildland–urban interface fires and major industrial fires were significantly higher.

A variety of projects were identified in order to improve system performance. The 16 different upgrades in 20 combinations are indicated in figure 8.2, which shows the resulting system reliability immediately following a Concord M6.5 event increasing qualitatively to the right. A cost–benefit analysis was then performed, incorporating the cost of these improvements versus the expected loss as a result of the earthquake (where loss includes direct damage, plus loss due to failure of the water service, primarily due to fire losses resulting from the earthquake). The total cost (i.e. the cost of improvements plus loss due to the earthquake) was higher for iterations 1–10, dipping to a low for iteration 17, the optimum (least total cost). This and several other near-optimum iterations all incorporated the ***backbone*** improvement, a carefully selected trunk line routing avoiding poor soils.

Emergency water supply and disaster vulnerability

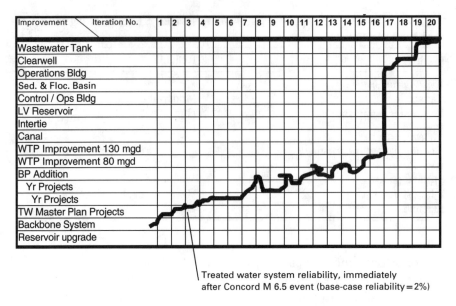

Fig. 8.2 **Iteration alternatives for CCWD cost–benefit analysis**

With heavy-wall pipe, special joints, and isolation valves, this routing has a high likelihood of being functional following a large earthquake and, because its routing has been chosen to be within 1 mile (1.6 km) of virtually all customers, it would be a reliable post-earthquake source of fire-fighting water for the District.

San Francisco

San Francisco, with a population of 700,000 and total consumption of about 300 mgd (1.2 million m³ per day), derives its water via the Hetch Hetchy system, which also serves 1 million people on the San Francisco Peninsula. The Municipal Water Supply System (MWSS) distribution network within the City of San Francisco contains 1,900 km of pipe, of which about 89 per cent is cast and ductile iron. The eastern portion of the city, which contains the central business district (CBD), also has significant areas of liquefiable soils (6.7 per cent of total length). The MWSS configuration is such that the CBD, served almost exclusively from the University Mound terminal reservoir, will have its feeder mains interdicted by breaks in these liquefiable soils (as happened in 1906). The MWSS was chosen as the case study for the ATC 25-1 Model Methodology (Scawthorn, Khater, and Rojahn, 1993). This study analysed the MWSS for an M8.3 San Andreas event (similar to that in 1906), and determined that the MWSS would sustain about 925 breaks in toto (0.48 breaks/km). It is of interest to note that the MWSS sustained about 150 breaks in the 1989 Loma Prieta earthquake, of which 95 per cent were due to liquefaction. The ATC 25-1 study examined in detail the time required for restoration and the economic impacts resulting from these breaks, finding overall that significant portions of the system would be out of service for about two weeks, that various parts of the city (disaggregated by zip code) would have 20–40 per cent loss of function at one month after the earthquake, and therefore would sustain indirect economic losses ranging from 7 per cent to 14 per cent of the monthly gross product of the zip code. The effects of fire following earthquake were not considered in this analysis, but it was obvious that the MWSS was quite unreliable as a post-earthquake emergency water supply. This conclusion had been obvious following the 1906 earthquake, which had led to the construction of the Auxiliary Water Supply System (AWSS).

The AWSS is a special high-pressure system maintained by the San Francisco Fire Department. Completely separate from the MWSS, the AWSS comprises approximately 129 miles of cast iron and ductile iron pipe serving approximately 1,500 hydrants. The AWSS does not serve buildings, is not a potable supply, and is especially concentrated in the high-value north-east quadrant of the city (the only built-up area when the system was constructed; see fig. 8.3). It has numerous redundant features, including several alternative water sources (two pump stations on San Francisco Bay capable of delivering 10,000 gpm, a 10 million gallon reservoir, 172 cisterns of 75,000 gallons each, and fireboat manifolds).

Analysis of the AWSS in the late 1980s indicated that the system, although very robust and reliable compared with ordinary water systems, still had vulnerabilities, owing to its interconnectedness and the time required to isolate breaks, and that the system also did not protect the entire city, which had grown considerably since its initial construction (Scawthorn et al., 1988). This was confirmed soon thereafter, when the AWSS sustained only a few breaks in the 1989 Loma Prieta earthquake, but lost all pressure in the Lower Zone for several hours until breaks could be isolated. As a result, the existing AWSS is installing special radio-controlled seismic motor-operated valves (MOV) at a number of locations. These valves close automat-

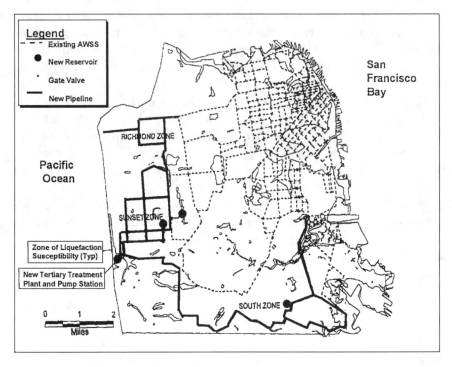

Fig. 8.3 **San Francisco Auxiliary Water Supply System, including dual-use addition in western portion of San Francisco (Source: Scawthorn et al., 1995)**

ically in an earthquake, but are radio-controlled with stand-alone power supplies and can be opened and closed from a central and backup locations. As a result, the AWSS will be very reliable.

In order to extend the AWSS protection to outlying parts of San Francisco, the Department of Public Works is cooperating with the San Francisco Fire Department to build a dual-use fire protection/reclaimed water (RCW) system (fig. 8.3). The design of this system was reliability based, and began with a detailed analysis of three different demand conditions: (a) an M8.3 event on the San Andreas fault, similar to the 1906 event, with associated fires following the earthquake; (b) potential non-earthquake-related large fires; and (c) irrigation demands. Large non-earthquake fire demand was based on an analysis of approximately 50,000 fires occurring during the 1980–1993 period, including a sub-set of almost 400 large fires. The procedure employed for the reliability analysis is shown in figure 8.1 above. Results are indicated in table 8.2.

Table 8.2 **Reliability analysis: San Francisco reclaimed water/fire protection dual-use addition**

Reservoir	Max. size		Required capacity						Reliability[a]	
			RCW		Large fires		Earthquake fires (Avg/Prob Max)			
	10^6 gal.	10^3 m^3	10^6 gal.	10^3 m^3	10^6 gal.	10^3 m^3	10^6 gal.	10^3 m^3		
Richmond	5	20	2.8	1.2	2.8	11.2	12/22	48/88	RCW	VH
									Fire	H
									FFE	NA
Sunset	15	60	4.0	6.0	4.5	18.0	9/14	36/56	RCW	VH
									Fire	VH
									FFE	MH
South	20	80	4.2	6.8	5.0	20.0	16/24	64/96	RCW	VH
									Fire	VH
									FFE	MH

a. Reliability is defined as the likelihood that the system will be able to meet performance criteria: VH = very high (>>90%); H = high (about 90%); MH = moderately high (50% ~ 90%). RCW = reclaimed water; Large fire = non-earthquake fire emergency; FFE = fire following M8.3 earthquake; NA = not available.

Vancouver

Vancouver (population 400,000), like San Francisco, is located on the Pacific "ring of fire" and has a significant history of earthquakes. A study by KJC Engineers et al. (1993) for example employed an M7 earthquake within 50 km of Vancouver as its design basis or 475-year event. This is comparable to but larger than the 1989 Loma Prieta event, which caused significant liquefaction and a number of fires in San Francisco. In contrast, however, the Vancouver region and the City of Vancouver in particular have an especially vulnerable water supply (as confirmed by the study by Ballantyne et al., 1996). The City of Vancouver receives most of its water from the Capilano (via the 1st Narrows Crossing) and Seymour (via the 2nd Narrows Crossing) reservoirs. Both of these pipelines cross potentially liquefiable alluvial deposits on the north shore of Burrard Inlet, so there is a high probability of failure of water delivery to Vancouver (an additional source is Coquitlam, but this has also been judged likely to fail in the same event). In 1986 a break occurred at the 1st Narrows Crossing, resulting in difficulties in closing the Capilano transmission line (owing to the butterfly valves), and water restrictions in Vancouver during the repair period.

Within Vancouver, storage consists of one Greater Vancouver Water District reservoir, Little Mountain Reservoir (LMR), dedicated for use by the City of Vancouver, with two other reservoirs (Kersland and Sasamat) used by GVWD for operational capacity. LMR has a capacity of 30 million Imperial gallons (Ig), insufficient for one day's supply during peak summer months. Should an earthquake result in significant pipe breakage, Little Mountain Reservoir would likely be quickly drained. Approximately 75 per cent of Vancouver's current water mains are cast-iron pipe, which is relatively weak and brittle and prone to break under earthquake shaking (EQE, 1990).

Regarding fires following earthquakes in Vancouver, no systematic engineering assessment has been performed comparable to that for San Francisco, Los Angeles, and other North American cities (Scawthorn and Khater, 1994). One study (EQE, 1990) briefly considered the problem, concluding that: "Emergency water demand is estimated to be 20,000 to 30,000 Igpm for fire fighting for several hours following an earthquake, or approximately 7 million Imperial gallons."

In recognition of the relatively high risk in the Lower Mainland, the City of Vancouver has followed the example of San Francisco and

Emergency water supply and disaster vulnerability

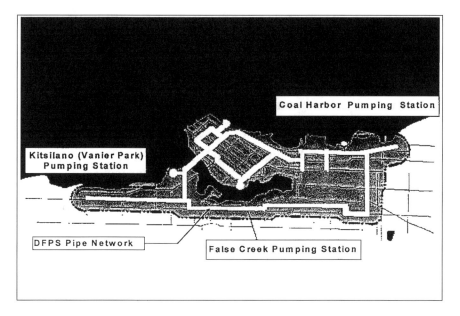

Fig. 8.4 **Dedicated Fire Protection System, City of Vancouver (Source: Scawthorn, 1996b)**

embarked on construction of a high-pressure Dedicated Fire Protection System (DFPS), which will contribute enormously to the protection of the high-value parts of the city, and in general all parts of the city (Mickelson and Moore, 1995). The DFPS is envisioned eventually to comprise four 10,000 Igpm pump stations (located at Coal Harbor, False Creek, the south shore of English Bay, and west of downtown, near the 2nd Narrows Crossing), connected by looped, seismically resistant, high-pressure high-volume pipe (fig. 8.4). The False Creek DFPS pump station was commissioned in September 1995, and, as of this writing, the Coal Harbor DFPS pump station was complete. Plans are under development for the construction of the first phase of a pipeline connecting these two pump stations.

System reliability analyses for the relatively small DFPS network were performed, and found that reliability of the DFPS pump stations was a key component of overall reliability. In order to assess the pump station reliability, pump station component relationships were mapped in a Fault Tree Diagram and combined with component fragilities to assess overall pump station reliability.

Component seismic fragilities are defined as the probability of loss of function of the component, given a measure of seismic ground motion (such as peak ground acceleration). Loss of function of the component refers to the earthquake preventing proper functioning of the component; for example, a pump not being able to function because the electrical relays in the motor control cabinet have been damaged by earthquake vibrations. Component seismic fragilities can be determined via several methods, including:

- **testing**, such as on a shake table – this is an expensive procedure, and does not replicate field installations and the effect of other nearby structures and equipment on the component;
- **analytical**, that is, calculations based on inherent factors of safety – this is also relatively expensive, because it requires the development of modes of failure and considerable materials and other data for the analyst;
- **experience based**, that is, statistical or bounds based on the performance of similar components in actual earthquakes – although requiring extensive data collection, this has proved to be a very powerful approach (Scawthorn et al., 1992).

The actual component fragilities employed in the DFPS pump station reliability assessment were derived from the EQE Earthquake Experience Data Base, the world's largest data base of earthquake experience performance, with over 250,000 data items collected from 65 earthquakes since 1971.

Station reliability analysis was performed using standard methods of fault tree analysis, and proved to be a valuable tool in preliminary design, serving to provide a framework for determining the number and layout of pumps, drives, backup power, and other station components.

Zaire refugee camps

As noted above, in July 1994, as a result of a civil war in Rwanda, 1.5 to 2 million Rwandan refugees crossed the border into Zaire. As noted by Biswas and Tortajada Quiroz (1995):

The influx of Hutu tribespeople from Rwanda into Zaire is unprecedented. Never before in history has so many people moved to a new country within such a short time, nor has such an economically disadvantaged nation ever been forced to take in so many refugees.

Some of the refugee camps had no satisfactory drinking water available at the site. The water and sanitation conditions at the vari-

Emergency water supply and disaster vulnerability

ous camps were such that a cholera epidemic occurred, resulting in daily mortality rates of 6,500 persons per day.

In order to alleviate this situation, an aid mission including a specialist technical contractor from the United States (Portable Water Supply System Co. Ltd, of Redwood City, CA) arrived in Goma on 26 July and established a potable water supply within 24 hours. This conveyed water from a lake source about 10 km away in high volume via a large-diameter hose system. A schematic of the water distribution system set up by PWSS is shown in figure 8.5. The system consisted of the following components:

- A specialty floating pump, termed the Hydrosub, manufactured in the Netherlands and capable of furnishing approximately 4,000 litres per minute at 10 bar. The Hydrosub has several attractive features including: a relatively lightweight portable pumphead (approximate weight, 50 kg), which floats on the surface and can draft water and other liquids from the surface (i.e. it is also useful for chemical spills) in shallow waters (e.g. 50 cm); a hydraulic power supply delivered via hydraulic lines up to 100 m long, driven remotely from a diesel engine, as shown in figure 8.5; portability – the entire Hydrosub including diesel engine, pump head, and fuel supply is trailerable.
- A large-diameter hose (LDH), in this case 125 mm in diameter, capable of carrying service pressures in excess of 20 bars. Approximately 2 km of LDH can be carried on a vehicle that is portable by air.
- Portable hydrants, consisting of a special casting with four gate valves. Portable hydrants are interspersed every several lengths of LDH, so that branch LDH or other hose can be coupled into the supply main, and water taken off at will.
- Pressure-reducing valves (not shown in fig. 8.5), modelled after the Gleeson valve design, which are lightweight and allow precise pressure setting irrespective of flow volume.

Final delivery was 1.6 million litres per day. This water was chlorinated at 5 ppm using portable chlorinators, achieving a residual dose of 0.75 to 1.0 ppm at delivery points. Water distribution was via taps branched off the LDH using manifolds, as well as via 100 tanker loads per day distributing in the camps.

As indicated in figure 8.6, the incidence of cholera, dysentery, and dehydration dropped from about 6,000 cases per day to fewer than 1,000 within 11 days of this system's installation. Mortality, too, dropped precipitously following the delivery of potable water (fig. 8.7).

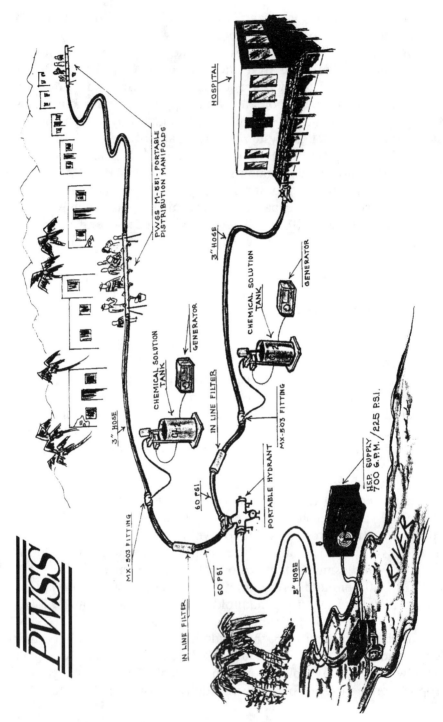

Fig. 8.5 **Rural water purification distribution system (Source: PWSS Ltd, Inc.)**

Emergency water supply and disaster vulnerability

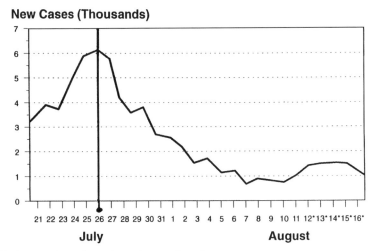

*Estimates based on dysentery outbreak.

Fig. 8.6 **Incidence of cholera, dysentery, and dehydration in refugee camps in Goma, Zaire, July–August 1994 (Note: water supply operation by US Army and PWSS began on 26 July. Source: UNHCR Medical J5-HOC GOMA)**

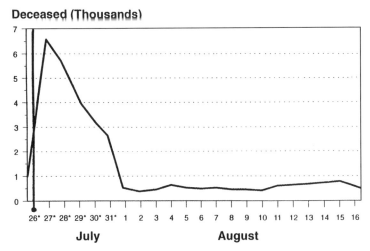

*Rate determined by nonsystematic means.

Fig. 8.7 **Daily mortality in refugee camps in Goma, Zaire, July–August 1994 (Note: water supply operation by US Army and PWSS began on 26 July. Source: UNHCR Medical J5-HOC GOMA)**

219

Emergency water supply and disaster vulnerability

The concept of an emergency water supply system

The foregoing sections have illustrated the following points:
1. urban water systems have *not* performed well in a variety of recent natural hazards;
2. methods are readily available to assess the vulnerability of these systems to natural hazards and to develop cost-effective mitigation measures;
3. these methods are readily applied, as has been illustrated by several recent examples.

However, the widespread application of these methods to achieve a reduction in vulnerabilities of urban water supply systems is unlikely to occur in the near or mid-term. This is for a number of reasons, the primary one probably being lack of resources.

On the other hand, water systems are rarely likely to become disrupted by natural hazards for extended periods. The most important damage is usually to transmission piping, pumps, water treatment plants, and other facilities that can usually be repaired within a few days to weeks. Longer-term repairs to perhaps thousands of breaks in distribution piping may require months, but local potable water distribution can be effected by tankers or above-ground temporary mains.

None the less, the examples in this chapter have shown that even short-term disruption to water systems can be sufficient to result in catastrophic loss of life and property, as in Tokyo in 1923 or in the Zairean refugee camps. Although the need for water is relatively immediate and on-going, it does not necessarily have to be of the same quality, in the same quantities, or use the same delivery mechanisms as permanent installations. This leads to the recognition that perhaps two kinds of emergency water delivery need exist:
- within **minutes**, e.g. for immediate fire suppression or hazardous materials response;
- within **hours**, e.g. for a potable supply.

Immediate need (for fire suppression, etc.) is typically not widespread. Rather it is confined to specific areas of a region, which can be identified prior to an event using standard planning methods. In these areas, large volumes of water may be required (for fire suppression for example) relative to the size of the area, but these volumes of water will still be small relative to the regional needs of a population. The problem is often not the supply, but *transmission* to a specific site. The chronic need for a potable supply on the other hand requires larger volumes. However, the real problem is *distribution* of these volumes, in a capillary fashion, to populations measured perhaps in millions. In this case, some compromises can be made, in that tanker trucks or faucet manifolds at local distribution points will suffice.

Consideration of these two different needs, and experience in several of the case studies discussed above, lead to the realization that both needs can be served by the same basic system, similar to that deployed in the 1991 East Bay Hills fire and in Zaire. This is a large-diameter hose (LDH) system together with necessary appurtenances, such as pumps, portable hydrants, valves, etc. In urban regions where the need is likely to be immediate, such as large cities with wooden buildings in the United States or Japan, LDH emergency water supply systems could be developed and deployed within cities, perhaps by fire departments themselves, so as to be available within minutes. In other urban regions, where the need is measured in hours, LDH emergency water supply systems could be located on perhaps a national or regional basis, so that they can be deployed by air, arriving perhaps the next day (as occurred in the Zaire example). What emerges from this is a concept of a global corps of LDH emergency water supply systems, perhaps several dozen in number, located in areas where the need for an emergency water supply would typically be immediate. When disasters struck in other regions, where the need could be met within say 24 hours, a few of the LDH emergency water supply systems could be deployed from each nation or region to the disaster-stricken area. This would constitute a major resource for furnishing potable water supply to the population until satisfactory repairs could be effected to the municipal system.

The need for an immediate water supply exists at a minimum in conflagration-prone areas, such as in Japan, the United States, and New Zealand. If one LDH emergency water supply unit per 1 million inhabitants were established in large urban regions, this would mean about 20 in the Tokyo region, 10 in the Osaka–Kobe–Kyoto region, 10 in the Los Angeles region, 6 in the San Francisco region, and so on. As an example, if a major earthquake struck Taipei, then the deployment of perhaps two units from each of the larger regions (e.g. Tokyo, Osaka, Los Angeles) would mean that within 24 hours some 12 km of LDH above-ground water mains could be laid in Taipei for potable water supply transmission and distribution. This would be sufficient to provide survival water rations to the entire population.

This leads to the following proposal.

Proposal

This paper proposes that the United Nations or other international relief agency study the concept of an international system or corps of LDH emergency water supply systems, located in a number of nations or regions around the globe, to determine its potential for effective disaster response. It is not proposed that the emergency water supply corps be an agency of the United Nations or other international relief agency. Rather, it is suggested that the United Nations or other agency could provide a coordination role, by:
1. effectively and clearly establishing the need for such a corps,
2. developing appropriate standards for equipment, response, and other parameters,
3. identifying nations where it would be appropriate to locate one or several LDH emergency water supply systems, and
4. developing support within national fire or other agencies in these nations, for their participation in the corps.

This proposal is conceptual in nature and requires additional definition, but it provides the seed of an idea that I hope can be grown to fruition.

Concluding remarks

Water systems have been damaged and lost functionality in several recent disasters. To avoid this, several water supply projects have employed reliability methods to determine system configuration and even details of hardware. Table 8.3 summarizes the reliability aspects of these projects vis-à-vis the reliability process defined above. Although reliability analysis of pipe networks has been available for some time, these projects were innovative in rationally defining system performance criteria, quantifying these criteria via empirical and analytical methods, and incorporating these criteria in an overall reliability and optimization framework. Without quantification of performance and direct linkage to system design via reliability and cost–benefit analysis, system design would have been based on prevailing industry standards, with only a loose fit to the system and its region's specific needs. The result was a clearer understanding of system needs and a more robust and cost-effective design.

It is clearly necessary for urban regions to implement this process. However, a number of impediments, such as lack of resources, preclude implementation of this process for many regions. To fill this

Table 8.3 **Summary of reliability aspects of case-study water supply system projects**

Reliability process	Contra Costa Water District	San Francisco MWSS	San Francisco AWSS/RCW	Vancouver DFPS
Performance criteria	Stress events (earthquake, large fires, power loss, etc.)	Non-major emergency	Earthquake, large fire, irrigation	Fire following earthquake
(example)	M6.5 earthquake causes 40 fires, requires 97,000 gallons/min	Potable and commercial use	M8.3 earthquake requires 12 ~ 22 million gallons	182,000 litres/min (after limited analysis)
Alternative designs	20 combinations of 16 different upgrades	Not considered	Various pipe routes, reservoir configurations	Limited pipe routes (pump station sites constrained)
Reliability analysis	Serviceability (Monte Carlo)	Serviceability (Monte Carlo)	Serviceability (Monte Carlo)	Serviceability (Monte Carlo)
Cost–benefit analysis	Yes	No	Limited	Limited
Final design	Backbone system and other measures	None (case study only)	New pipe, reservoirs, etc.	2 pump stations built; pipe network under design

need, it is proposed that the United Nations or other international relief agency investigate the concept of an international cooperative *corps of emergency water supply systems*, with the aim of coordinating multinational development and participation in such a corps.

Notes

1. See, for example, the excellent series of articles on mega-city growth and the future in Fuchs et al. (1994).
2. Unless otherwise noted, M refers to earthquake magnitude as measured on the moment magnitude scale.

References

Avila, E. A. and Chan, T. K. 1995. "Reliability Criteria for Capital Project Planning," *Proc. Fourth US Conf. on Lifeline Earthquake Engineering*, Am. Soc. Civil Engineers, San Francisco, CA, August.

Ballantyne, D. 1995. "Water and Wastewater Systems," in *The Hanshin-Awaji Earthquake of January 17, 1995, Performance of Lifelines*, NCEER Report No. 95-0015, National Center for Earthquake Engineering Research, State University of New York at Buffalo.

Ballantyne, D., Heubach, W., and Archibald, P. 1996. "Earthquake Vulnerability of the Greater Vancouver Water District's Pipeline System," *Proc. Pan Pacific Hazards '96 Conference*, Vancouver, B.C., July.

Biswas, A. K. and Tortajada Quiroz, C. 1995. "Rwandan Refugees and the Environment in Zaire," ECODECISION, Spring.

――― 1996. "Environment Impacts of Refugees: A Case Study," *Impact Assessment*, Vol. 14, No. 1, March.

Eguchi, R. T. and Chung, R. 1995. "Performance of Lifelines during the January 17, 1994 Northridge Earthquake," *Proc. Fourth US Conf. on Lifeline Earthquake Engineering*, Am. Soc. Civil Engineers, San Francisco, CA, August.

Eidinger, J., Maison, B., and Lau, B. 1995. "East Bay Municipal Utility District Water Distribution Damage in 1989 Loma Prieta Earthquake," *Proc. Fourth US Conf. on Lifeline Earthquake Engineering*, Am. Soc. Civil Engineers, San Francisco, CA, August.

EQE Engineering and Design. 1990. *Recommendations for Improvements Required for Emergency Water Supply*, report prepared for City Engineering Dept., City of Vancouver, under sub-contract to CH2M-Hill., San Francisco, CA.

Fuchs, R. J., Brennan, E., Chamie, J., Lo, F.-C., and Uitto, J. I. (eds.). 1994. *Mega-City Growth and the Future*, United Nations University Press, Tokyo.

KJC Engineers and EQE Engineering and Design. 1993. A *Lifeline Study of the Regional Water Distribution System*, Final Report prepared for the Greater Vancouver Water District, December.

Mickelson, P. and Moore, D. E. 1995. "City of Vancouver Dedicated Fire Protection System Underground Piping Design Considerations," *Proc. Fourth US Conf. on Lifeline Earthquake Engineering*, Am. Soc. Civil Engineers, San Francisco, CA, August.

Scawthorn, C. 1987. *Fire Following Earthquake – Estimates of the Conflagration Risk to Insured Property in Greater Los Angeles and San Francisco*, prepared for the All-Industry Research Advisory Council, Oak Park, IL.

——— 1996a. "Fire Following the Northridge and Kobe Earthquakes," paper presented at UJNR Panel on Fire Research and Safety, National Institute of Standards and Technology, Gaithersburg, MD, March.

——— 1996b. "Reliability-based Design of Water Supply Systems," paper presented at 6th Japan–US Workshop on Earthquake Resistant Design of Lifeline Facilities and Countermeasures Against Liquefaction, Waseda University, Tokyo, June.

Scawthorn, C. and Blackburn, Frank T. 1990. "Performance of the San Francisco Auxiliary and Portable Water Supply Systems in the 17 October 1989 Loma Prieta Earthquake," *Proceedings 4th U.S. National Conference on Earthquake Engineering*, Palm Springs, CA.

Scawthorn, C. and Khater, M. 1994. "Fires Caused by Earthquakes: A Greater Threat Than Many Realize," *NFPA Journal*, May/June, pp. 82–86.

Scawthorn, C. and O'Rourke, T. D. 1989. "Effects of Ground Failure on Water Supply and Fire Following Earthquake: The 1906 San Francisco Earthquake," *Proceedings, 2nd U.S.–Japan Workshop on Large Ground Deformation*, July, Buffalo.

Scawthorn, C., Bouhafs, M., and Blackburn, F. T. 1988. "Demand and Provision for Post-earthquake Emergency Services: Case Study of San Francisco Fire Department," *Proc. 9th World Conference of Earthquake Engineering*, Tokyo and Kyoto.

Scawthorn, C., Khater, M., and Rojahn, C. 1993. "A Model Methodology for Assessment of Seismic Vulnerability and Impact of Disruption of Water Supply Systems," *Proc. 1993 National Earthquake Conference*, Memphis, TN.

Scawthorn, C., Swan, S. W., Hamburger, R. O., and Hom, S. 1992. "Building Life-Safety Systems & Post-Earthquake Reliability: Overview of Codes and Current Practice," *Proc. of Seminar and Workshop on Seismic Design and Performance of Equipment and Nonstructural Elements in Buildings and Industrial Structures*, Applied Technology Council, Redwood City, CA.

Scawthorn, C., Odeh, D. J., Khater, M., Blackburn, F., and Kubick, K. 1995. "Reliability Analysis of a Dual Use Fire Protection/Reclaimed Water System, San Francisco CA," *Proc. Fourth US Conf. on Lifeline Earthquake Engineering*, Am. Soc. Civil Engineers, San Francisco, CA, August.

Scawthorn, C., Khater, M., Isenberg, J., Lund, L., Larsen, T., and Shinozuka, M. 1990. "Lifelines Performance during the October 17, 1989 Loma Prieta Earthquake," UJNR Panel on Wind and Seismic Effects, National Institute of Standards and Technology, Gaithersburg, MD, April.

UNHCR. n.d. *Manual for Environmental Surveys and Studies. Technical Support Document for Interim Guidelines for Environment-Sensitive Management of Refugee Programmes*, United Nations High Commissioner for Refugees, Geneva.

Yanev, P. I. and Scawthorn, C. 1993. *Hokkaido Nansei-oki, Japan Earthquake of July 12, 1993*, NCEER Report 93-0023, National Center for Earthquake Engineering Research, State University of New York at Buffalo.

9

Conclusions

Juha I. Uitto and Asit K. Biswas

The world has seen an unprecedented growth in water consumption during the twentieth century. Much of this increase can be attributed directly to population growth, which has been especially pronounced in the developing countries since the Second World War. The world population has more than doubled since the 1950s.

This is, however, only part of the story. Per capita water consumption has gone up dramatically, as well. This has been mainly due to increased demand for agricultural water, although industrialization, urbanization, and economic growth have contributed. As people get richer they increase their water consumption. This lifestyle change is now affecting large populations in the developing world. Economic growth in the South is likely to lead to massive increases in water consumption in coming years. Asit Biswas has shown in chapter 1 the impacts of population growth, urbanization, and economic development on the consumption of water on a global level. Although urban consumption still represents a fairly small percentage of the total, especially compared with agriculture, its share is going to rise in the future.

The problems of urban water concern quality as much as quantity. Urban areas produce huge amounts of sewage – from both households and industries – and this causes problems for both human and ecosystem health. Water-related diseases are prevalent, especially in the poorer parts of developing country cities. Similarly, providing sufficient amounts of water to the growing urban centres is in itself a challenge. It is thus obvious that solutions need to be sought on multiple fronts involving all stakeholders.

In the past, emphasis has been placed on infrastructure development. Development of proper and sufficient infrastructure for urban water supply and sewage is, of course, necessary. However, more attention must be paid to the soft side of water management, including demand management. Improvements in policies, pricing, and operation and maintenance are equally important.

The efficiency of water-related infrastructure is often appallingly low. In chapter 4, Rajendra Sagane highlighted the problems facing four Indian mega-cities where demand for water is outstripping supply. He concluded that much of the blame can be placed on poor operation and maintenance of the systems. The emphasis has been on the construction of costly infrastructure, while operation and management of the existing systems have been neglected. There is now an urgent need to move towards more efficient management.

The example of Japan serves to demonstrate what careful management can contribute to the efficiency of urban water supply. In chapter 2, Yutaka Takahasi outlined the historical development of the water supply and sewage works in Tokyo. The capital of Japan has experienced remarkable growth during the twentieth century and is now arguably the largest urban conglomeration in the world (with some 31 million people in eight adjacent cities). Losses from the urban water supply system of Tokyo have been reduced from 80 per cent in 1945 to the current remarkably low 9.9 per cent. Infrastructure development in Japan has now reached maturity. The problems today are different. It is now important to focus on providing good-quality water at the least possible cost. Technological developments can play an important role in this, but cannot alone solve the problems.

Another Japanese case, as described by Masahisa Nakamura (chap. 3), highlights the need to balance the requirements of urban areas for water with the ecosystem demands in the region surrounding the cities. Lake Biwa is the largest lake in Japan and important for the water supply for the second-largest urban conglomeration in the country – the Kansai area, which consists of the cities of Osaka, Kyoto, Nagoya, and Kobe. The case study of the Lake Biwa–Yodo

River system shows how the individual municipal water systems have evolved separately on the basis of the agendas of the various municipalities in the basin. A holistic approach to integrated watershed management has been lacking. Consequently, policy has been fragmented and implementation has been incremental, causing problems with the economics of water use.

Similarly, environmental considerations and restoration are becoming increasingly important with regard to water resources management and quality. In Japan, the development of water resources has been seen in the context of overall economic development, with an emphasis on river development, dam construction, diversion channels, etc. In the future, however, it will be important to incorporate environmental considerations and conservation into water resources development plans.

The Lake Biwa Comprehensive Development Project (LBCDP) took 25 years to complete. This was achieved during a time of very rapid economic growth in Japan when environmental concerns were not considered as front-running issues. Similar projects today and in less wealthy countries would take much longer to realize.

A case in point may be Mexico City, a sprawling metropolitan area of 25 million people, as described by Cecilia Tortajada-Quiroz in chapter 5. Population growth remains high and water demand is increasing rapidly. To cater for the growing demand, Mexico has embarked on projects that involve long-distance water transfers from basins located far from the city. As a consequence, costs are escalating rapidly. Similarly, supplying water to the urban giant is taking a toll on the environment in the surrounding areas. The situation will not be sustainable in the longer term.

The conclusions emerging from this volume are clear: providing water and sanitation for urban areas in the twenty-first century will pose a major challenge for humankind in the years to come. This challenge will require new solutions and policies to be developed. In particular, given the escalating conflicts in water resources development and increasing water demand in the developing countries, wastewater reclamation and reuse will in the future need to be seen as increasingly important options in the sustainable development of urban water resources. According to Takashi Asano (chap. 6), demand for water in the city is an issue of the reliability as well as the dependability of water supply. Reused wastewater can be used for purposes where quality requirements are not that high: flushing toilets, washing cars, gardening, etc.

Pollution control is also essential and should be strictly enforced. Industries that are polluting should be made responsible for paying for the costs of the clean-up.

Another dimension to the issue is the vulnerability of water supply systems to natural and manmade disasters. Many of the world's urban centres are in locations susceptible to hazards. It is, for example, estimated that by the year 2000 three-quarters of the population of the United States will live within 15 km of either coast. The same trend is evident everywhere in the world. Coasts in general are exposed to climatic hazards, such as typhoons and hurricanes, that can disrupt water supply and sewerage systems. Incidents of flooding can have devastating effects on urban areas. Furthermore, events such as earthquakes can lead to the loss of water service, which is one of the essential lifelines of the city. The disaster vulnerability of water systems in urban areas is tragically evident from Charles Scawthorn's chapter in this book (chap. 8).

An important consideration in making urban water supplies more sustainable is the financial mechanism governing water supply and use. Water is still seen as a free commodity, and water prices to consumers seldom – if ever – reflect the true cost of water. This leads to waste and no incentive to conserve water. Authorities – whether national, city, or local – should adjust the price of water to reflect the true cost of bringing it to the consumer. All phases of the water supply must be taken into account, including the costs of operation and maintenance of the system. Instead of taxing consumers directly, indirect taxation could be adopted. However, in some cases subsidies still need to be used.

One way to improve efficiency is privatization of waterworks in cities. The utilization of private sector services is certainly very important. In chapter 7, Walter Stottmann has provided a review of the lessons to be learned from the operations of the World Bank, a leading proponent and financier of private sector involvement in the water sector. His examples clearly demonstrate the benefits from subjecting the construction, operation, and management of water supply and sewerage services to competition by private suppliers. However, it is clear that privatization should not be seen as a panacea. Private sector involvement may be essential for improving the efficiency of waterworks, but in many cases the private sector does not have the capacity to handle the services on its own. Therefore, private–public partnerships are needed.

Developing countries, in particular, have limited financial resources. External assistance in the form of aid or loans is vital for developing countries because water supply schemes for mega-cities are very costly. There is, however, a need to reassess the policies of international bilateral and multilateral development agencies. For example, the World Bank today prefers rehabilitating existing waterworks rather than building new ones, or water demand management rather than financing another plant. The problem is that this approach often goes against national policies and against the interests of those in power. Politicians like to build new plants for prestige reasons; rehabilitation and efficiency improvements are seen as much less attractive.

The industrialized countries need to rethink the focus and priorities for their official development assistance programmes in the sector. Thus far the emphasis has been on the construction of new infrastructure projects. What is needed now is a long-term vision for broad-based cooperation in the water sector that places more emphasis on the "software" aspects, such as capacity building and institutional strengthening. This cooperation needs to include both resource flows from the North to the South and technology and knowledge transfers.

It will also be important to change national policy environments to facilitate efficient water management in a holistic manner. Public awareness needs to be raised to support these processes. Educational strategies, including non-formal educational programmes on environmental and social issues, are needed. Educating the next generation to become responsible water users is an important task for the future.

The world faces an urgent water crisis of dimensions that no earlier generation has had to face. These problems must be solved lest there be a major human tragedy, especially in the developing countries. However, the problems can be solved with correctly focused investment, technology, and management. What is needed to achieve this is political will and active collaboration between North and South, and East and West. If countries can initiate urgent action, there is reason to be cautiously optimistic about the future.

Contributors

Takashi Asano
Department of Civil and Environmental Engineering, University of California at Davis, CA, USA

Asit K. Biswas
President, Third World Centre for Water Management, Mexico City, Mexico

Masahisa Nakamura
Director, Lake Biwa Research Institute, Shiga, Japan

Rajendra Sagane
Member Secretary, Maharashtra Jeevan Pradhikaran, Mumbai, India

Charles Scawthorn
Senior Vice President, EQE International, Tokyo, Japan, and Oakland, USA

Walter Stottmann
Water and Sanitation Sector Leader, The World Bank, Washington, D.C., USA

Yutaka Takahasi
Professor Emeritus, Department of Civil Engineering, Shibaura Institute of Technology, Tokyo, Japan

Cecilia Tortajada-Quiroz
Third World Centre for Water Management, Mexico City, Mexico

Juha I. Uitto
Monitoring and Evaluation Specialist, Global Environment Facility, Washington, D.C., USA

Index

activated carbon (water treatment)
  process 35, *144*
advisers, private sector 194–195
aerobic biological treatment process
  143
Africa
  population growth *4*
  urbanization xii
Agenda 21 xvi
agriculture
  pollution by 68–69
  water for 2, 99, 100
    in competition with domestic usage
      12
    in India 99, 100
    in Mexico 112, *113*
air stripping (water treatment) process
  144
algal bloom 76
Algiers, water project costs 14
alternative designs for reliable water
  supplies 201, *223*
Amman [Jordan], water project
  costs 14
Ara River [Japan], canal connecting to
  Tone River 30, 31
Argentina
  freight rail concessions 192, 193
  private sector water concessions *161*,
    164, 181, 192
  regulatory system 175
Arizona [USA]
  groundwater recharge 146–147
  wastewater reclamation criteria 148
arrears of water charges 15, 95
Asaka Canal [Japan] 30
Asia
  population growth *4*
  urbanization xii
  water services in various cities
    compared 20
ATC 25-1 study 211
Australia
  private sector contracts *161*, 166
  regulatory system 175

Bangkok [Thailand]
  consumption per capita 19, *20*
  population growth *6*
  service indicators 20

Index

water service 16, 17, *40*
Basin Council for the Valley of Mexico 131–132
Beijing [China]
  consumption per capita *20*
  household water bills 15
  population 5
  service indicators *20*
  utility staff ratios 16
bids *see* private sector funding, bidding process
biochemical oxygen demand (BOD)
  Osaka City's discharge 61
  range in treated wastewater *140, 145–146*
biological nutrient removal process *143*
biological treatment processes *143*, 149
Biwa–Yodo water system [Japan] 47–49, 51–52, 227–228
  integrated watershed management 78–79
  organizations covering 78–79
  pollution of 35, 68–70, *71*
  upstream–downsteam relationships 65–68
  water resource development projects *53*
  see also Lake Biwa; Yodo River
Bombay [India]
  population trends xii, 5, *5*
  squatter settlements 9
  see also Mumbai
borewell supplies 91, 98
bottled water usage xii, 2, 18, 19
Buenos Aires [Argentina]
  regulatory system 175
  staff levels of public water utility 181
  water concession 164, 198
    bidding process 192
    staff reduction 181
    time required to prepare 195
build–operate–own (BOO) arrangements 165, *170*
build–operate–transfer (BOT)
  contracts 165–166
  bidding process 193
  duration of contract *161*, 165, *170*
  enabling conditions *186*

examples *161*, 166
information required for bid *186*, 187
key responsibilities *161*
level of investment *161, 170*
objectives *185*
and regulatory framework 171, *186*
risk analysis *161*, 182–183
variations 165
buildings, use of reclaimed wastewater 43, 45, 92, 131, *145*

Cairo [Egypt], population growth 6
Calcutta [India]
  consumption per capita *20*, 88, 101
  population 5, *88*, 101
  slum population 85, *88*, 101
  water supply 101–105
    availability (hours per day) *20*, 102
    leakage rates *20*, *88*, 89, 104
    planning for future needs 104–105
    problems of management 104
    production costs *20*
    repair of distribution network 105
    service indicators *20*, 21, *88*
    sources of water 101, *102, 103*
    supply capability *88, 102*
    tariffs 102, 104
California [USA]
  earthquakes 203–204
  fires 204–205
  groundwater recharge 146–147
  reclaimed water applications *147*, 212, *213*
  wastewater reclamation criteria 148, 149, *150*
  wastewater reclamation regulations 141
Cape Town [South Africa], water service *40*
capital cost
  desalination plant 93
  wastewater reclamation plant 150
chemical oxygen demand (COD), Lake Biwa [Japan] *74*
Chennai (Madras, India)
  population *88*, 105, 108
  slum population 85, *88*
  water availability in 12

233

Index

Chennai (Madras, India) (cont.)
  water management   108–109
  water supply   105–109
    future projects   106, 108
    leakage rates   *88*, 89
    sources of water   105–106, *107*
    supply capability   *88*, 105
    utility staff ratios   16
Chile
  private sector contracts   161–162, *161*, 166
  regulatory system   175
China, private sector contracts   *161*, 166
closed-loop recycling systems   138–139
coliform count, range in treated wastewater   *140*, *145–146*, *150*
combined sewer systems   62, 127
  designs to deal with storms   62, 70, 128
competition
  allowed/exerted by regulator   168, 196
  factors affecting   166–167
  for market   167–168
  in market   168–169, 196
competitive bidding   188–189, 190
competitive negotiations   189
component seismic fragilities   215–216
  determination of   216
Conakry [Guinea], regulatory system   175
concession contracts   163–164
  bidding process   193
  duration of contract   *161*, 163, *170*
  enabling conditions   186
  examples   *161*
  information required for bid   *186*, 187
  key responsibilities   *161*
  level of investment   *161*, *170*
  objectives   185
  and regulatory framework   169, 171, 175, *186*
  risk analysis   *161*, 182–183
  time required to prepare   195
Concord Fault [USA], earthquake   209
connection levels/percentage
  in Indian cities   *20*, 98
  in Japanese cities   *26*, *35*
  in Mexico   113–114, *114*, 119
  various cities compared   *20*
connectivity analysis   207
conservation measures
  in India   93–94, 100–101, 108–109, 110
  in Mexico City   131
  in Tokyo   37–38, 44
constraints on water availability   10–22
  economic costs   13–14
  environmental and health issues   18–19
  financing   14–16
  management constraints   16–18
  mind-sets of utility managers   19, 21–22
  scarcity of water   11–13
construction costs, water-storage projects   13, *14*
consumption per capita
  India   *20*, *88*, 91
  Japan   *39*
  Mexico   112, 117, *118*, 120
  various cities compared   19, *20*, *39*, *40*, *88*
Contra Costa Water District system [USA]   208–210, *223*
contracts, private sector
  bidding for   187–194
  enabling conditions   *186*
  managing   195–197
  pre-contract analysis   177–183
  renegotiation of   196
  types   160–166
cost–benefit analysis   201, 208
  in case studies   *223*
  example   *210*
Côte d'Ivoire, private sector contracts   *161*, 163
Cutzamala Aquaférico [Mexico]   126
Cutzamala Macrocircuit [Mexico]   126
  investment cost   126
    recovery of   132
  supply capability   126
Cutzamala river system [Mexico]   116, 119
Cutzamala System [Mexico]   123–126
  area affected by construction   124

Index

energy requirements to run   125
Environmental Impact Statement for fourth stage   123, 124, 125
hydropower capacity   124
investment cost   124, 125
operational running costs   125
social projects   124–125
Czech Republic, private sector contracts   *161*, 163

Delhi [India]
  consumption per capita   19, *20*, *88*
  electricity-to-water bill ratios   15
  household water tariff   98
  plans for future   101
  population density   96
  population trends   *6*, *88*, 96
  slum population   *85*, *88*
  water supply   95–101
    availability (hours per day)   19, *20*
    distribution pipe length   98
    leakage rates   *20*, *88*, 89
    problems in management   98–99
    production costs   *20*, 98
    scope for improvement   99–101
    service indicators   *20*, 21, *88*
    sources of water   96–97
    supply capability   *88*
    tariffs   98
Denver Summit of Eight   xvii
Des Moines Water Works [USA], flood   205
desalination of sea water   43, 92–93
  capital cost of plant   93
design–build–operate (DBO) arrangements   165
Detroit [USA], water service   *40*
Dhaka [Bangladesh]
  financial management of water utilities   15
  population   *5*, *6*
  water project costs   *14*
direct filtration
  compared with Title 22 Process   141
  life-cycle costs   *151*
direct negotiations (private sector)   189–190

disaster vulnerability of water supplies   200, 202–206, 229
disasters
  performance of water supply systems in   202–206
    earthquakes   203–204
    fires   204–205
    floods   205
    refugee camps   205–206
diseases
  waterborne   xiii, 98, 129
  in Zairean refugee camps   217, *219*
disinfection (water treatment)   *143*, 217
distribution pipe length
  Delhi   98
  Japanese cities   *39*
  Mexico City   116
  various cities compared   *40*
divestiture arrangements   *161*, 166
  bidding process   193
  enabling conditions   *186*
  objectives   *185*
  and regulatory framework   166, 169, *186*
domestic usage of water, in competition with agricultural usage   12, 99
drinking water
  availability
    in developing economies   xii, xv, 2, 18
    in western developed economies   2
  health risks   18–19
  quality monitoring parameters   139, *140*
droughts   2

earthquakes
  damage caused by
    in California [USA]   203–204
    in Japan   25, 26, 41, 63, 204
economic contribution of urban areas   8
  in India   84
  in Japan   49–50
economic costs, as constraint on water availability   13–14
Edo City *see* Tokyo
Edo River [Japan]   29, 35
electricity generation, water for   2

235

Index

electricity-to-water bill ratios 15
emergency water supply system 220–221
   proposal for international corps 221–222, 224
   in Tokyo 42
   in Zaire 217
Endhó Dam [Mexico] 129, 130
England & Wales
   privatization of public utilities *161*, 166
      competition aspects 168
      price cap regulation 173
      regulatory system 175
EQE Earthquake Experience Data Base 216
Europe, population growth *4*
eutrophication
   control of 72
   of Lake Biwa [Japan] 73–75
   of Osaka Bay [Japan] 72
evaporation control 93, 100
experience-based methods 216

Faisalabad [Pakistan], electricity-to-water bill ratios 15
fault tree analysis 216
filtration (water treatment) process 124, *143*
financial management of utilities 15–16, 94–95, 158, 181–182
financing constraints 14–15, 104
fires 204–205
flocculation/precipitation (water treatment) process 124, *144*
flood control systems
   Japanese cities 45, 48–49, 55
   Mexico City 127–128, 130
floods 2, 44, 205
Florida [USA]
   reclaimed water applications *147*
   wastewater reclamation criteria 148
France
   private sector contracts *161*, 163, 164
   privately managed utilities 159
Fukuoka [Japan], water service *39*

Ganges River [India] 102, *103*
Gdansk [Poland], joint venture lease contract 165
Geneva [Switzerland], water service *40*
Grand Forks metropolitan area [USA], flood 205
groundwater extraction, effects 36, 60, 120
groundwater recharge
   reclaimed wastewater used 138, *138*
      examples 36–37, 43, 129, 130, 146–147
   treatment goals *146*
Guinea
   private sector contracts *161*, 163
   regulatory system 175

Hanoi [Vietnam], household water bills 15
Hanshin–Awaji earthquake [Japan] 41, 63, 204
Hanshin Industrial Complex/Zone [Japan] 50, 56
Hanshin Water Supply Corporation [Japan] 59
health risks 18–19, 98, 129, 148
   measurement of pathogenic organisms *140*
Hiroshima [Japan], water service *39*
Hong Kong [China]
   consumption per capita *20*
   household water bills 15
   service indicators *20*
   water utility 16
household water bills 15, 104
household water tariffs
   Indian cities 94, 95, 98
   Japanese cities *39*
   Mexico City 121, *121*
Howrah [India] 101
   sources of water 102, *103*
   see also Calcutta
Hyderabad, water project costs 14
Hydrosub floating pump 217

India
   bottled water usage 18, 19

Index

mega-cities
  population trends   85, *88*
  preferential treatment   85–87
  slum populations   85, *88*
  strategies needed to achieve sustainable supplies   109–111
  water supply examples   *88*, 89–109
  water supply problems   87, 89
  population trends   85, *88*
industrial development, water for   2
industrial water supply systems   2, 36, 60, *61*
  recycled wastewater used   138, 142, 147
    example applications   *146*
    in India   92
    treatment goals   *146*
informal settlements   9–10
  low priority by governments   9–10
  see also squatter settlements
infrastructure development   7–8
integrated water resource management   78–79
intermittent water supplies   19
  examples   *20*, 91
  storage facilities for   21
international corps of emergency water supply systems   222, 224
International Network on Water, Environment and Health (INWEH)   xvii
International Water Supply and Sanitation Decade (IDWSSD)   10
  effects   10–11
ion exchange (water treatment) process   *144*
irrigation
  wastewater used for   142, *145*
    in India   92, 99–100
    in Mexico   129–130, 131
    quality guidelines   *150*
    treatment goals   *145*
  see also agriculture, water for

Jakarta [Indonesia], population trends   xii, 5, *5*

Japan
  earthquakes   25, 26, 41, 63, 204
  infrastructure development   7–8
  Overseas Economic Cooperation Fund   109
  reclaimed water applications   36–37, 43, *147*
    factors affecting   147
  water service compared for various cities   39
  see also Kawasaki; Kobe; Kyoto; Osaka; Tokyo; Yokohama
joint venture leases or concessions   164–165
Jokaso system   65

Kansai Metropolitan Region [Japan]   49–51, 81[1], 227
  emerging issues   77–80
  pollution control and wastewater management   60–65
  population   47, *49*
  population density   56
  water metabolism   56–65
  water quality issues   65–75
  water resources   51–56
  water supplies   57–60
  see also Kobe, Kyoto, Osaka
Kanto earthquake [Japan]   25, 26
Karachi [Pakistan]
  consumption per capita   *20*
  electricity-to-water bill ratios   15
  financial management of water utilities   15
  population   5, *6*
    growth   xii, 5, *6*, 9
  service indicators   *20*, 21
  squatter settlements   9
Kathmandu [Nepal], electricity-to-water bill ratios   15
Kawasaki [Japan], water service   28, *39*
Kita-Kyushu [Japan], water service   *39*
Kobe [Japan]   *48*, 50
  earthquake damage   50–51
  sewerage system   63
  water service   *39*

237

Index

Kuala Lumpur [Malaysia]
  consumption per capita 20
  service indicators 20
  water utility 16, 17
Kusaki Dam [Japan] 25, *34*
Kyoto [Japan] *48*, 50
  sewerage system 62–63
  waterworks 39, 58
    balance between cost and quality 68

Lagos [Nigeria], population trends xii, 5, *5*
Lake Biwa Comprehensive Development Project (LBCDP–Japan) 47–49, 55–56, 78, 228
  factors affecting 75, 77
  financial arrangements 66
  new phase of management 80
Lake Biwa [Japan] *48*, 227
  catchment area 51
  pollution of 35, 73–75
  population served by *49*, 57
  size (area and volume) 52
  sustainable water use 78
  wastewater system to protect 63–65, 77
  water quality
    new phase in management 80
    trends 74
  see also Biwa–Yodo water system
Lake Biwa–Yodo River Water Quality Conservation Organization 79
large-diameter hose (LDH) systems 217, *218*
  applications
    in international emergency water supply systems 221, 222
    in refugee camps 217
Latin America and Caribbean region
  population growth *4*
  sewage/wastewater treatment 7
  urbanization xii
leak detection 38, 93, 100
leakage rates 21, 22, 157
  in Indian cities 20, *88*, 89
  in Mexican cities 120, *121*
  reasons for lack of action 21–22

Tokyo waterworks 28–29, 38, 227
  various cities compared 20, *88*
lease contracts 163
  bidding process 193
  duration of contract *161*, *170*
  enabling conditions *186*
  examples *161*
  information required for bid *186*, 187
  key responsibilities *161*
  level of investment *161*, *170*
  objectives *185*
  and regulatory framework 175, *186*
  risk analysis *161*, 182–183
Lerma–Balsas river system [Mexico] 116, 119
Lima [Peru], water project costs *14*
lime (water treatment) process *144*
Loma Prieta [USA] earthquake, damage caused by 203, 211
London [UK]
  population growth *6*
  reservoir capacity 41
Los Angeles [USA]
  earthquake 203–204
  emergency water supply system 221
  population *5*

Macao, private sector contracts *161*, 164
Madras [India] *see* Chennai
Madras Metro Water Supply and Sewage Board (MMWSSB) 108, 161
Malaysia
  private sector contracts *161*, 164, 166
  see also Kuala Lumpur
Malé [Maldives]
  consumption per capita 19, *20*, 21
  service indicators 20
management contracts 162–163
  bidding process 193
  duration of contract *161*, *170*
  enabling conditions *186*
  examples *161*
  information required for bid *186*, 187
  key responsibilities *161*
  level of investment *161*, *170*
  monitoring of performance 171, *186*
  objectives *185*

Index

risk analysis *161*, 182
time required to prepare 195
management inefficiencies 15–17, 157–158
Manila [Philippines]
  population growth *6*
  water concession 164, 168, 195
mega-cities
  definition (by size) 4, 135
  growth rates 5, 6
  loss of water supply due to natural disasters 200
  number in world 135, 157
  population data listed 5, 38, *88*
  see also urbanization
Melbourne [Australia], competition among service providers 168
membrane filtration (water treatment) *144*
metering of water supplies 15, 110, 122
Mexico
  irrigation by wastewater 129–130
  leakage rates 120, *121*
  macro water projects 122–128
  National Water Commission (CNA) 118, 119
    plan for Mexico Basin/Valley 131
    social projects 124, 125
  population 112
  rainfall data 112
  water availability 113–114, *114*
  water demand 114
  water use by sector 112, *113*
Mexico Basin/Valley, precipitation/evapotranspiration data 116
Mexico City 115–117
  consumption per capita 117, *118*, 120
  population 5, 7, 117, *118*
  population density 117
  population growth rate 117, 128
  sewerage system 117, 127–128
  subsidence 120, 127
    effect on water/sewerage systems 120, 127, 128
  wastewater
    reuse of 129–130
    treatment of 117
    volume 117, 127

water project costs *14*, 122–123
water supply 2, 116–117
  distribution pipe length 116
  leakage rates 120
  management contracts 162
  production costs 121, 125, 131
  storage capacity 116
  tariffs 121, *121*
  see also Cutzamala System
water utility 17, 118
Mexico City Metropolitan Zone (ZMCM) 117
  characteristics *118*
  constraints on water resources management 128–131
  contamination of aquifer 129
  water consumption 118
  water demand 117–119
  water supply
    problems 119–122
    sources of water 119
Mezquital Valley [Mexico], irrigation with wastewater 129–130
Monte Carlo techniques 207, *223*
Monterey Regional Water Pollution Control Agency [USA], treatment costs for water reuse *151*
Monterey Wastewater Reclamation Study 141
multilateral development institutions, support of private involvement 159–160
Mumbai (Bombay, India)
  consumption per capita 19, *20*, *88*
  household water bills 15
  population trends xii, 5, *88*, 89
  slum population 85, *88*
  water supply 89–95
    alternative sources of water 92–93
    availability (hours per day) 19, *20*, 91
    dams/reservoirs (current and proposed) *90*, 95
    financial management 15, 94–95
    leakage rates *20*, *88*, 89
    problems in management 91–92
    production costs *20*
    scope for improvement 92–94

239

Index

Mumbai (Bombay, India) (cont.)
   service indicators 20, 88
   sources of water 89–90, 90
   supply capability 88, 91
   tariffs 94, 95
   utility staff ratios 16
   see also Bombay
Musashi Canal [Japan] 25, 30, 31

Nagoya [Japan], water service 39
Naramata Dam [Japan] 25, 34
nematodes, quality criteria in treated wastewater 149, 150
New York [USA]
   population 5, 6
   reservoir capacity 41
New Zealand, private sector contracts 161, 166
Nobidome Canal [Japan] 36
North America
   population growth 4
   see also USA
Northridge [USA], earthquake 203–204
nutrients
   in Mexican irrigation water 130
   range in treated wastewater 140

Oakland Hills [USA], fires 204
Oaxaca [Mexico], system losses 21
Oceania, population growth 4
Ogouchi Dam [Japan] 25, 27–28
   drop in levels 29
   volume of reservoir 27
Okushiri [Japan], earthquake 204
Olympic Games, Tokyo [1964] 30
operational management of water utilities 15–16
organophosphoric acid triesters
   industrial uses 81–82[5]
   pollution by 69, 70
Osaka Bay [Japan] 48
   industrial developments 47, 50
   marine activities 52
   water quality 72–73
   see also Biwa–Yodo water system
Osaka [Japan] 48, 50
   industrial water supply systems 60, 61

sewerage system 62
waterworks 39, 59, 60
   balance between cost and quality 68
   pollution of water sources 35
Osaka–Kobe–Kyoto region
   emergency water supply system 221
   see also Kansai Metropolitan Region
oxidation (water treatment) pond 143

paddy field runoff, pollution caused by 69, 75
Paris [France]
   private service providers 168
   reservoir capacity 41
parks, irrigation by reclaimed wastewater 99, 145
pathogenic organisms, count range in treated wastewater 140, 145–146, 150
performance criteria, reliable water supply 201, 223
pesticides, pollution by 69, 71
Poland, private sector contracts 161, 165
political commitment to private sector ventures 185, 186
political interference in management of water utilities 17
pollution
   water supplies affected by
     in India 98
     in Japan 35, 68–72
Pomona Virus Study 141, 151
population density
   Delhi 96
   Mexico City 117
   Tokyo 38
population growth 3–4, 156–157
   Indian cities 88, 89, 96
   Tokyo 26
Portable Water Supply System (PWSS)
   water distribution system 217, 218
   use in Zairean refugee camps 217
pre-contract analysis (private sector) 177–183
   assessment of utility 177–178

Index

financial analysis   181–182
regulatory/institutional analysis   178–179
risk analysis   182–183
for smaller cities   183
stakeholder analysis   179–181
pre-qualification of (private sector) bidders   190–191
price regulation   172–173, 196
price control/cap   172–173
rate-of-return/profit control   172
prices, domestic water supply
changes by private sector   181–182, 184
India   94, 95, 98, 102, 104, 110
Japan   *39*
Mexico City   121, *121*
private sector funding   158–160, 229
bidding process
bid contents   192–193
complaints/appeals procedure   193–194
evaluation of bids   192–193
information available   *186*, 187–188
organizing the bidding   188–190
pre-bid contacts with bidders   191–192
pre-qualification of bidders   190–191
competition   167–169
duration of contract   *161, 170*
enabling conditions for various options   *186*
examples   *161*, 164, 181, 192, 198
key responsibilities of various options   *161*
objectives of various options   *185*
performance comparison of service providers   168
regulation   169–171
considerations in defining regulatory framework   171–176
types of participation   *161*
build–operate–transfer (BOT) contracts   *161*, 165–166
concessions   *161*, 163–164
divestiture   *161*, 166
hybrid arrangements   166
joint ventures   164–165
leases   *161*, 163
management contracts   *161*, 162–163
service contracts   160–162, *161*
private sector venture
managing the contract   195–197
contract renegotiation   196
maintaining competitive pressure   196
prevention of undue outside interference   196–197
managing the process   194–195
independent advisers   194–195
management unit   194
time requirements   195
preparing for   177–194
attractiveness of option to private sector   184–186
choosing among options   183–184
finding suitable partner   187–194
pre-contract analysis   177–183
privatization of utilities   166, 229
see also divestiture arrangements
production costs
Indian cities   *20*, 94, 98
Mexican cities   121, 125, 131
various cities compared   *20*
public taps   17–18
see also standpipe supplies

quality monitoring parameters, reclaimed wastewater   139, *140*

rain making by cloud seeding   93
Rakunan wastewater treatment plant [Japan]   63, *67*
Rakusei wastewater treatment plant [Japan]   63, *67*
reclaimed water
applications   142, 144–147, 212, 228
in India   92, 99–100
in Japan   36–37, 43, *147*
in Mexico   129–130, 131
in USA   *147*, 212, *213*
future developments   152–153

241

Index

reclaimed water (cont.)
  health and regulatory
    requirements   148–149, *150*
  in hydrological cycle   137–138
recycling of water   138
  in India   92, 93
  in Tokyo   45–46
red tide [phytoplankton]
  in Lake Biwa [Japan]   73, *76*
  in Osaka Bay [Japan]   72
refugee camps, drinking water
    availability   205–206, 216–219
regulatory framework   169–171
  advisers required   194
  considerations in defining   171–176
    accountability of regulator   175
    areas to be regulated   171–172
    discretion of regulator   173–174
    finding appropriate system   175–176, 184
    independence of regulator   174
    locus of regulation   173
    price regulation   172–173, 196
  effect on choice/design of private
    sector arrangement   178–179
rehabilitate–operate–transfer (ROT)
    arrangements   165
reliability analysis (for water supply)
    201, *202*, 206–208
  applications   208–219, *223*
    Contra Costa Water District system
      208–210, *223*
    San Francisco Auxiliary Water
      Supply System   211–212, *223*
    San Francisco Municipal Water
      Supply System   210–211, *223*
    San Francisco reclaimed water
      system   212, *213*, *223*
    Vancouver Dedicated Fire
      Protection System   214–216,
      223
reliable water supply, performance
    criteria for   201
reverse osmosis (water treatment)
    systems   93, *144*
risk analysis, in private sector
    participation   *161*, 182–183
Rome [Italy], water service   40

Rotterdam [Netherlands], water service
    40

Sagami River [Japan]   28, 29
salaries of water utility managers   16
San Andreas earthquake   211
San Francisco [USA]
  Auxiliary Water Supply System   203,
    211–212, *223*
  earthquakes   202, 203
  emergency water supply system   221
  reclaimed water/fire protection system
    212, *213*, *223*
  water supply system   203, 210–211,
    223
  reservoir capacity   41
sanitation   8
  in Japan   65, 68
  in Mexico   114
  number of people lacking access
    to   xii, xv, 8, 156
Santiago [Chile], water utility service
    contracts   161–162
São Paulo [Brazil], population   5
Sapporo [Japan], water service   *39*
scarcity of water resources   11–13, 44
seawater desalination   43, 92–93
sedimentation (water treatment) process
    124, *143*
Sendai [Japan], water service   *39*
Senegal, private sector contracts   *161*,
    163
Seoul [South Korea], population   5, 6
separate-sewer system   62, 63
  compared with combined system   62
service contracts   160–162
  bidding process   193
  duration of contract   161, *161*, *170*
  enabling conditions   *186*
  examples   *161*
  information required for bid   *186*,
    187
  key responsibilities   *161*
  level of investment   *161*, *170*
  monitoring of performance   171, *186*
  objectives   *185*
  risk analysis   *161*, 182
  time required to prepare   195

242

Index

service indicators, various cities compared 20, 35, 39, 88
serviceability analysis 207
Seto Inland Sea [Japan]
  environmental protection of 72
  see also Osaka Bay
sewage
  Lake Biwa watershed [Japan] 77
  treated
    use in India 92, 99
    use in Mexico 131
    use in Tokyo 36–37, 43
  treatment of xii, 7, 36
  see also wastewater
sewerage systems
  combined compared with separate sewer systems 62
  cost of upgrading US systems 136
  Japanese cities 36–37, 62–63
  Mexico City 127–128
Shanghai [China]
  financial management of water utilities 15
  population 5
Shenyang [China], water project costs 14
Shimokubo Dam [Japan] 25, 31, *34*
Singapore
  consumption per capita 20
  water service 16–17
    service indicators 20, *40*
    staff ratio 16
Slovenia, private sector contracts 166
slum populations, in Indian cities 85, *88*, 101
smaller cities, private sector ventures for 183
Spain, private sector contracts *161*, 163, 164
squatter settlements 9
  see also informal settlements
stabilization ponds 149, *150*
staff of utilities
  overstaffing 16, 157–158, 181
  as stakeholders *180*, 181
stakeholder analysis 179–181
  potential issues and policy responses *180*

standpipe supplies 91
  see also public taps
storage tanks, in houses 21
stress events, in reliability analysis 208
subsidence, causes 36, 60, 120
supply capability
  Indian cities *88*
  Japanese cities 26, 27, 39, *54*, *58*, 59
Surabaya [Indonesia], water project costs *14*
sustainable water use, Kansai area [Japan] 77–78
Sydney [Australia], regulatory system 175
system losses *see* leakage rates

Taipei [Taiwan], water utility 16
Tama River [Japan] 24
  water resources development 32, *34*
  dams 25, 29, 32, *34*
  water shortages 25, 26, 29
Tamagawa Canal [Japan] *25*, 36
tariffs *see* prices, domestic water supply
Tashkent [Uzbekistan]
  electricity-to-water bill ratios 15
  household water bills 15
Temascaltepec project [Mexic] 123
Tianjin [China]
  household water bills 15
  utility staff ratios 16
Title 22 Process [for wastewater treatment] 141
  treatment costs *151*
Toba wastewater treatment plant [Japan] 62, 67, 81[4]
Tokyo [Japan]
  earthquakes 25, 26, 41
    preparation for 41–42
  floods 44
  future water resource policies 43–44
  industrial water supply 36
  Kanda Canal 24
  Musashi Canal 25, 30, 31
  pollution of water supplies, actions taken 35, 41, 42
  population 5, 38, 44
  population density 38
  population growth xii, 5

243

Index

Tokyo [Japan] (cont.)
  recycling of water in 45–46
  river improvement work 45
  sewerage systems 36–37
  Shinjuku Suburbanization Plan *25*, 31
  urbanization effects 44–45
  utilization of treated sewage 36–37, 43
  "water conservation conscious society" approach 37–38, 44
  water shortages *25*, 26, 29–30
    supply plans to prevent 30, 40–41
  waterworks
    Asaka Purification Plant *25*, 31, *33*
    compared with other cities 38–40
    distribution pipe length 32, *35*, *39*, *40*
    future targets 40–43
    Higashi-Murayama Purification Plant *25*, 31, 32, *33*
    history 24–32
    Kanamachi Purification Plant *25*, *33*, 35, 42
    leakage rate 28–29, 38, 227
    Misato Purification Plant *25*, *33*, 35
    Misono Purification Plant *25*, *33*
    Nagasawa Purification Plant *25*, 27–28
    Ogouchi Dam project *25*, 27–28
    Ozaku Purification Plant *25*, *33*
    population served by 26, *35*, *39*, 40
    purification plants *25*, 32, *33*, 35
    reservoir capacity *34*, 41
    Sakai Purification Plant *25*, 32, *33*
    service indicators 26, *35*, *39*
    supply capability 26, 27, *35*, *39*
    Tama River system 24, 29, 32, *33*, *34*
    Tamagawa Purification Plant *25*, *33*
    Tone River system 29, 30–32, *33*, *34*
    war damage 28–29
    water resources feeding 32, *34*
    Yodobashi Purification Plant *25*–26, *25*, 31
  Waterworks Bureau 37, 38, 40

Tone Estuary Barrage [Japan] *25*, 31
Tone River [Japan] 29, 32
  water resources development 30–32
  dams *25*, 29, 31, 32, *34*
  purification plants *25*, 31, *33*
total organic carbon (TOC), range in treated wastewater *140*
total suspended solids (TSS), range in treated wastewater *140*
Trinidad & Tobago, management contracts *161*, 162
turbidity, range in treated wastewater *140*, *145–146*

unaccounted-for water
  in India *20*, *88*, 89, 157
  various cities compared *20*
  see also leakage rates
unions, as stakeholders 181
United Nations Conference on Environment and Development [Earth Summit] xvi
United Nations General Assembly Special Session Earth Summit+5 xvii
United Nations University, research and training programme xvii
urbanization
  effects on water circulation 4–10, 44–45
  global growth xii, xv, 4
  in India 85
  in Japan 44
  problems 8–9, 84
  rate of growth 5–7, 9, 85
  see also mega-cities
USA
  earthquakes 202, 203–204
  fires 204–205
  floods 205
  management contracts *161*
  privately managed utilities 159, 166
  sewerage systems, cost of upgrading 136
  wastewater reclamation criteria 148
  see also Arizona; California; Florida; New York

244

Index

utility companies
  financial management considerations 15–16, 94–95, 158, 181–182
  management inefficiencies 16, 157
  staff numbers/ratios 16, 157–158, 181

Vancouver [Canada], Dedicated Fire Protection System 214–216, *223*
Vienna [Austria], water service *40*

wastewater
  amount produced, in Mexico City 113, 117, 127
  reuse of
    applications 142, 144–147, 212, 228
    future developments 152–153
    health and regulatory requirements 148–149, *150*
    in hydrological cycle 137–138
    in India 92, 99–100
    in Japan 36–37, 43
    in Mexico 129–130, 131
    objectives 136
  treatment of 7, 12, 13–14
    direct filtration process 141
    economic costs 136, 149–152
    in India 92, 100
    in Japan 52, 59, 62–63
    in Mexico City 117, 131
    recycling facility 93, 100
    technologies involved 139–142, *143–144*
    Title 22 Process 141
  see also sewage
water availability
  constraints on 10–22
  developing compared with western economies 2
  in Mexico 113–114, *114*
  various cities compared *20*
  see also constraints on water availability

water resource planning
  in India 95, 101, 104–105, 108
  reclaimed wastewater in 154
water supply
  economic costs 13, *14*
  number of people lacking access to safe supply xii, xv, 156
water supply projects, costs xii, *14*
water transfer schemes, political aspects 1
western economies 1–2
World Bank-funded projects 90
World Health Organization (WHO), quality guidelines for reclaimed wastewater 149, *150*
world population growth 3

Yagisawa Dam [Japan] 25, 29, 31, *34*
Yamaguchi Reservoir [Japan] 25, *34*
Yamuna River [India] 96, *97*, 98
yardstick competition 168, 196
Yodo River [Japan] *48*
  allocation of water rights *54*, 59
  catchment area 51
  development projects *53*
  hydrographic data 54–55
  pollution of 35, 69, *70*, *71*
  upstream–downstream relationships 66–68
  see also Biwa–Yodo water system
Yodo River Water Pollution Control Consultative Association 78–79
Yodo River Water Quality Consultative Association 79
Yokohama [Japan], water service 25, *39*

Zaire refugee camps 205–206, 216–219
  emergency water distribution system used 217–219
  incidence of diseases 217, *219*
  mortality data 217, *219*

245

CL

363.
610
917
32
WAT

6000430289